NIETZSCHE

NIETZSCHE

A Re-examination

IRVING M. ZEITLIN

Polity Press

Copyright © Irving M. Zeitlin 1994
The right of Irving M. Zeitlin to be identified as author of
this work has been asserted in accordance with the
Copyright, Designs and Patents Act 1988.

First published 1994 by Polity Press
in association with Blackwell Publishers Ltd.
Reprinted 1995

Editorial office:
Polity Press
65 Bridge Street
Cambridge CB2 1UR, UK

Marketing and production:
Blackwell Publishers Ltd
108 Cowley Road
Oxford OX4 1JF, UK

Blackwell Publishers Inc.
238 Main Street
Cambridge, MA 02142, USA

ISBN 0 7456 1290 3
ISBN 0 7456 1291 1 (pbk)

A CIP catalogue record for this book is available
from the British Library and the Library of Congress.

Typeset in 11 on 13 pt Ehrhardt
by Best-set Typesetter, Ltd., Hong Kong
Printed in Great Britain by Hartnolls Ltd, Bodmin, Cornwall
This book is printed on acid-free paper.

Contents

For Rebekka, Jacob, Kayla,
Isaiah, and Albert

Preface

No thinker in the history of modern philosophy has launched as radical an assault upon Western values as has Friedrich Nietzsche (1844–1900). Therein lies his importance as a philosopher – in challenging the religio-moral and intellectual foundations of Western society. On the whole, his challenge may be characterized as negative and critical, in that it aims to tear down the old table of values. Nietzsche's negative influence has left its mark not only on contemporary philosophy, social theory, literary criticism, and other academic areas, but on attitudes in everyday life as well.

What I have tried to accomplish in this book is a careful examination of Nietzsche's writings with the aim of laying bare what I regard as the problems, ambiguities, and defects in his work. Nietzsche, as we shall see, rejects both reason and metaphysics, thus leaving only taste as the criterion by which to choose between moralities, socio-political systems, and other human products and values. For Nietzsche, there are only moralities but no Morality, no trans-historical or universal truths. And since Nietzsche views reason and dialectics as manifestations of the "slave morality," and refuses to acknowledge the efficacy of reason in the pursuit of truth, the character of his utterances is essentially assertive. Nietzsche thus denies that a philosopher's primary responsibility is to provide sound arguments for what he or she believes. Nietzsche admires and prefers the "master morality," but this is, for him, a purely personal, "aesthetic" preference. Hence, the manifold and diverse values one finds in any complex society are, for Nietzsche, relative and ideological in the strict sense.

Although there may be some overlap with other studies in my exposition of Nietzsche's philosophy, I believe my work also contains distinc-

tive elements. For example, I give considerable attention to Nietzsche's sociological or social-psychological theory of *ressentiment* and the inversion of values. Nietzsche himself provides only the briefest outline of this theory with regard to the two classical cases with which he is most concerned, the Jews and the Greeks. I therefore develop and apply his theory in chapters 4 and 5, clarifying his meaning and illustrating the sociological fruitfulness of the theory. However, I also hasten to remind the reader that the social origin of an idea or value has no necessary implications for its validity.

Another distinctive element may be found in the discussion of Socrates and the proto-Nietzscheans (chapter 7). As the heart of Nietzsche's philosophy was anticipated by Callicles and Thrasymachus, I thought it would be interesting to hear how Socrates grapples with their arguments based upon the state of nature. Similarly, the excursus on Max Stirner and Karl Marx (chapter 8) enables us to see how Marx – a materialist like Nietzsche – counters the views of the boldest of Nietzsche's precursors in the modern era.

Like Callicles before him, Nietzsche bases his critique of democracy on what he perceives as the rule of nature in which the strong dominate the weak. He believed he was following Darwin in this respect but failed to realize that his understanding of Darwin's theory was one-sided and erroneous. The consequences of Nietzsche's error are discussed in "Darwin *contra* Nietzsche" (chapter 9).

Nietzsche was born in 1844 and died in 1900 – or for all practical purposes in 1890, since he spent the last ten years of his life totally incapacitated by his insanity. And although the issues he raised permits us to regard him as a twentieth-century philosopher, there is a sense in which his philosophy bears the stamp of the European nineteenth century – a century which, when compared with the twentieth, appears rather peaceful and civilized. If, therefore, Nietzsche had lived to see the totalitarian, genocidal regimes of the twentieth century, one wonders whether he would have proclaimed "God is dead!" as loudly and consistently as he did while ignoring the dangers of that proclamation. It was Dostoevsky who so brilliantly and dramatically addressed the question Nietzsche effectively ignored: "If God is dead, is everything permitted?" This is the subject-matter of what I call "Dostoevsky's Challenge" (chapter 12).

Finally in the light of my criticisms of Nietzsche's protagonist, Zarathustra, and the extreme poverty of his affirmations, I confront this spokesman for the "master morality," this yearner for the coming of the superman, with the Hebrew prophets of social justice, the classical representatives of the "slave morality." In that way we see clearly the choice that lies before us.

1

Introduction

The mature Nietzsche, as is well known, looked upon Christianity as a major source of the decadent and anti-life outlook of the West. Although we shall make no attempt to explain Nietzsche's philosophical ideas psychologically, it is noteworthy that he had spent the first five-and-a-half years of his childhood in a parsonage, and his subsequent childhood years in a home characterized by the kind of piety one might expect from the nature of his family background.[1] His father and his grandfathers on both sides had been Lutheran ministers, and indeed Nietzsche was the heir of a long line of Lutheran pastors reaching back to the early seventeenth century. Little wonder that early interpreters of Nietzsche rarely resisted the temptation of treating his mature philosophy simply as a violent repudiation of his religious upbringing. In any event, this upbringing together with his later theological studies imparted to him a firsthand knowledge of Scripture, both the Old and New Testaments.

Nietzsche ceased to be a believing Christian while still in his teens. In 1862, in an essay called "Fate and History," he set down the grounds for his doubts, stating that history and science are the only reliable means of pursuing truth and knowledge. It was no later than that date that Nietzsche abandoned his faith and adopted an extreme skepticism in which he proposed that an openness to divergent views is itself desirable. "Strife," he wrote, "is the perpetual food of the soul."[2] "Strife" is a translation of *Kampf* which alternates with *Krieg* and *Streit*, the German words Nietzsche employs throughout his mature writings to denote a state of active and aggressive struggle as opposed to peace and repose.

When Nietzsche enrolled in the university in Bonn, he eventually chose to study classical philology under the tutelage of the distinguished Friedrich Ritschl, who was greatly impressed with the young man's brilliance. Yielding to the powerful convention among university students, Nietzsche joined a dueling fraternity and promptly acquired a dueling scar, the prestigious sign of noble manliness. Manliness presupposed another kind of experience: on a brief trip to Cologne, a cab driver brought the young Nietzsche to a brothel where it is almost certain he contracted syphilis, the cause of the insanity to which Nietzsche fell victim during the last ten years of his life.

It was during his stay at Bonn that Nietzsche abandoned the study of theology, a discipline, he decided, devoted to the investigation of a primitive superstition, namely Christianity. He now declared himself to be a "free thinker"; but, in contrast to his atheistic predecessors from the time of the eighteenth-century Enlightenment, Nietzsche held that "freedom" meant not only throwing off a yoke but taking on a new and heavier one in its place. He had no patience for the so-called "liberal-minded" unbeliever who thought he could deny the existence of a divine Lawgiver but nevertheless accept the validity of the Law handed down by human beings. Nietzsche now began to sense that, insofar as God had ceased to exist as a reality for humanity, life as such was deprived of its traditional meaning. In such circumstances, humanity was liable to disintegrate and degenerate under the impact of its basic meaninglessness.

Although Nietzsche distinguished himself early as a brilliant young philologist, he held a rather modest view of the importance of that subject and often even deprecated it as the study of dead books. That attitude turned him away from philology and toward philosophy, where he discovered Schopenhauer and was for a time dazzled by him. At about the same time (1866) Nietzsche read Friedrich Albert Lange's *History of Materialism*, a work that influenced him profoundly, pushing him further in the direction of philosophical materialism. By now Nietzsche had left Bonn and followed Ritschl to Leipzig, where the young classical philologist so impressed Ritschl that the latter successfully recommended him for a vacant professorship at Basel University when Nietzsche was only 24 years of age. Ritschl then also expedited for Nietzsche his doctorate at Leipzig without an examination or any other formality. In November 1866 the University, at Ritschl's urging, designated Diogenes Laertius as the subject for a philological essay contest, and the prize was awarded to Nietzsche.

The Leipzig years (1865–9) also witnessed the emergence in Nietzsche's writings of at least two notable ideas which eventually contributed to the formation of his mature views. Following an intensive study of Homer and Hesiod and the role of the mythical "contest" in Greek culture, Nietzsche began to recognize just how central the concept of *agon*, or competition, was in the development of Greek culture.

Nietzsche remained at Basel for 10 years and when he left in 1879, at the age of 34, he had never been away from the classroom environment for more than a few months at a time. It was during the Basel years that he had met Richard Wagner, whom he idolized for a time as a creative genius of the "higher type." The notion of a "higher type" becomes in due course a chief element of Nietzsche's theory that Christianity and other ascetic, repressive, and enervating facets of Western culture cater to the weak and mediocre, thus obstructing the emergence of "higher specimens". For Nietzsche, the primary aim of a healthy and robust culture ought to be the fostering and nurturing of higher specimens of all sorts, an aim to which all else ought to be subordinated. It seems that what at first impressed Nietzsche most about Wagner, besides his bold experimentalism in music, was that he appeared to be the precise opposite of the bourgeois man for whom Nietzsche had acquired so much disdain. Wagner was a cultural rebel, an out-and-out Bohemian. As R. J. Hollingdale has observed, "From *Tannhäuser* onwards his operas reveal an unmixed contempt for normal standards of behavior and there is no Wagnerian hero who does not flout them."[3]

Wagner himself also broke standards, most notably in his outlandish and conspicuous dress and in his Bayreuth project, which was widely regarded as the undertaking of a megalomaniac for whom the existing opera houses were inferior and inadequate. Nietzsche's temporary intoxication with Wagner was strengthened by their joint admiration of Schopenhauer; but when Nietzsche broke away from Wagner in 1876 he also parted ways with the philosopher. Hollingdale has suggested that the fundamental difference between Nietzsche's mature theory and that of Schopenhauer is that Nietzsche had by this time rejected metaphysics in all its forms; but one needs to add that Schopenhauer's "will to live" is the precise opposite of Nietzsche's "will to power." Schopenhauer had proposed that the cause of suffering is intensity of will; the less we will, the less we shall suffer. He reaches this notion through love, which is always sympathy for the pain of others. Inspired by the Nirvana myth, Schopenhauer notes that sympathy can go so far as to take on the suffering of the whole world. In the good man, however, knowledge of the whole quiets all volition, and his will turns away from life and denies his own nature. The good man will practice complete chastity, voluntary poverty, fasting, and self-torment. Nietzsche, in contrast, rejects and despises asceticism. Nietzsche now saw himself as a fully consistent materialist whose "will to power" was an induction from observed phenomena, not a metaphysical postulate. Hollingdale acknowledges, however, that Nietzsche occasionally speaks of the "will to power" and the "eternal recurrence" as if these concepts were ultimate realities, a problem to which we shall return. On the whole, however, it is true that Nietzsche's thought-experiments in the form of aphorisms are secular. The most famous formulation of his "this-worldly" orientation is, of course, "God is dead." This is intended to include everything that has ever been subsumed in the concept of "God": other worlds, ultimate realities, "things in themselves," and valetudinarian "wills to live" of the kind Schopenhauer had proposed.

It was during his Leipzig and Basel years that Nietzsche had occupied himself with the peripheral figures of Schopenhauer and Wagner, whose thoughts belong, strictly speaking, to the nineteenth century. Nietzsche's mature thought, in contrast, is in the mainstream of Western philosophy; and though he died in the year 1900, he must be considered a twentieth-century philosopher because he raises key issues with which every thinking person ought to be concerned, the decline and loss of faith and the moral crisis that this implies. As Nietzsche's mature philosophy bears no trace of the influence of either Wagner or Schopenhauer, we can turn our attention to the several problem areas with which Nietzsche was preoccupied and which formed the central questions he addressed in his mature works: the nature of the Greek foundations of Western society, the "inversion of values", the nature and consequences of the Christian faith, and the implications of Darwin's theory of evolution by means of natural selection. In the present Introduction, we shall merely touch upon these problem areas, reserving for the body of this work a fuller exposition of Nietzsche's views and a critical response to them.

By the time Nietzsche had carefully read and pondered *The Origin of Species* and *The Descent of Man*, the validity of Darwin's theory appeared to him to have been demonstrated. Nietzsche accepted the fundamentals of the Darwinian thesis that humanity had evolved from earlier animal forms in a purely naturalistic manner, through chance and accident. Earlier evolutionary theories had still left open the possibility of a "purpose" in evolution; but Darwin had mobilized massive evidence in support of his view that "higher" animals and humans could have come into being entirely by fortuitous variations in individuals. Before Darwin had presented his convincing hypothesis, it was not too difficult to discern some directing agency in the unfolding of natural and, for some, even historical events. After Darwin, however, that became increasingly difficult. The need for a conscious, creative principle, force, or being seemed unnecessary, since what had formerly appeared as order could now be explained as random change. Hence, for Nietzsche, natural selection was a process free of metaphysical implication. Nietzsche now denied the existence of order in the universe, with the possible exception of the astral constellation in which we live, where a temporary order made possible the formation of organic nature. "The entire character of the world," he wrote, "is . . . in all eternity chaos, not in the sense of an absence of necessity, but of an absence of order, arrangement [*Gliederung*], form, beauty, wisdom, and whatever other terms there are for our aesthetic anthropomorphisms. . . . Let us beware of saying that there are laws in nature. There are only necessities: there is nobody who commands, nobody who obeys, nobody who trespasses" (*The Gay Science*, aphorism 109).[4] It is in this general context that we encounter the first occurrence of Nietzsche's most famous epigram: "God is dead: but given the way of humanity, his shadow will remain on the walls of caves for thousands of years. And we – we still have to conquer his shadow as well" (108).

Nietzsche's view of the world as chaotic was certainly reinforced by his reading of Darwin, and the chaotic nature of the universe remained a basic element of his philosophical outlook. The Darwinian theory complemented and confirmed a view of reality which Nietzsche had

already begun to form in his youth. From the time of his reading of
F. A. Lange's *History of Materialism*, Nietzsche had come to regard all
metaphysical ideas as mere ideas and nothing more. There was no such
thing as a supersensible reality with which humans could somehow get in
touch; and if earlier conceptions of biological evolution had still allowed
for the role of a divine agency in earthly matters, Darwin had put an end
to that once and for all. A conscious, directing agency was an unnecessary
hypothesis.

For Nietzsche, then, the logical consequences of Darwin's theory
were no less than revolutionary: God was no more than an idea in the
minds of human beings. Any "higher" attributes one may discern in
human beings are, in reality, attributes which they had acquired in the
course of their descent from other animals. Human beings possessed no
means of communicating with any so-called transcendental power, and
they were no different, fundamentally, from any other creature. And
if God was a mere idea, then it certainly could not give meaning to
the universe. Darwinism therefore implied for Nietzsche that the
planet Earth was devoid of any transcendental meaning. Nietzsche thus
regarded his own era as "nihilistic": all traditional values and meanings
had ceased to make sense, and philosophy was in a state of crisis, faced as
it was with an inherently meaningless universe. The various solutions
offered from the time of Plato were inadequate.

Nietzsche's intimate knowledge of Greek culture helped him develop
at least one idea with which to begin his project of providing secular
guidelines for a new and meaningful outlook in life. From the time of his
first book, *The Birth of Tragedy*, he continued to propose that Greek
cultural development, the Greek creative genius, was bound up with the
Apollinian-Dionysian duality, involving perpetual strife between the two
principles. The driving force behind the culture of Hellas was the contest, *agon*, the striving to surpass. The creative force is passion, but a
passion harnessed and directed. Dionysus is the explosive, ungoverned
force of creation while Apollo is the power that governs him. Is there
then a vital connection, Nietzsche now asked himself, between the
Darwinian view of intraspecies behavior and the genius of the Greeks?
Interpreting Darwin in his own distinctive way, Nietzsche proposed that
human qualities are of a twofold character, manifesting, to be sure, a
capacity for high and noble powers, but also for cruel, murderous, and

destructive ones. The Greeks certainly demonstrated a capacity for brutality, and yet they were also creative and humane; they were the inventors of philosophy, science, and drama. The Greeks were not simply beautiful and creative children, as some earlier scholars had portrayed them; they were a cruel, savage, and warlike people who constructed an extraordinarily valuable culture by governing and redirecting their passions and impulses. In this light it is clear that it was not Wagner and Schopenhauer but Darwin and the Greeks who were the starting point of Nietzsche's mature philosophy. In *The Birth of Tragedy*, Nietzsche also examines the role of Socrates, who is driven neither by Dionysus nor Apollo, but by something "new" – reason and dialectics – which represses the Dionysian passions and gradually causes the deterioration of Greek art and drama. In his first book, as in his later works, one discerns in Nietzsche a highly critical attitude towards the Platonic Socrates. This is a problem to which we shall need to return.

If for Nietzsche the universe was chaos and the traditional metaphysical meanings imposed on that chaos were useless for the provision of meaning, then what was needed was a new, secular, and truly convincing organization of the chaos. One of his chief early ideas in this regard seems to be Apollo's victory over, or control of, Dionysus. In his mature philosophy, however, Nietzsche attempts to go beyond this, and his organizing of the chaos leads to the chapter on "self-overcoming" in *Thus Spoke Zarathustra*, where the central Nietzschean idea "will to power" is first described:

> And life itself told me this secret: "Behold," it said, "I am *that which must always overcome itself*. . . . There is much that life admires more than life itself; but out of that very admiration speaks the *will to power*." That is what life once taught me; and with that I shall yet solve the riddle of your heart, you who are wisest. . . .
>
> And whoever would be a creator in good and evil, verily, he must first be a destroyer who breaks values. Thus the highest evil belongs to the highest goodness: but that is creative.

The single goal of humanity must be the creation of its highest specimens, the superman (*Übermensch*). Zarathustra spoke thus to the people:

I teach you the superman. Man is something that shall be overcome. What have you done to overcome him?. . . . The superman is the meaning of the earth. Let your will assert: the superman *shall be* the meaning of the earth! I urge you my brothers, *remain faithful to the earth*, and believe not those who speak to you of otherworldly hopes! Poison-mixers they are, whether they know it or not. (Part I, 3)

Nietzsche never defines the term "power," nor does he let us know, precisely, the nature of the superman's mission. At times self-mastery appears to be simply the means of achieving the highest aesthetic goals; but at other times self-mastery appears to be political – the means by which the "higher types" will dominate the "herd." Does Nietzsche write from a strictly aesthetic standpoint, as some scholars have argued, or from a political standpoint as well? And if from the latter, has he faced the Hobbesian problem of what occurs when two or more individuals desire the same apparent good, which nevertheless they cannot both enjoy? These questions will demand our attention in the following chapters.

Meanwhile, however, we need to say a word about Nietzsche's style of thinking and writing. Nietzsche was a "peripatetic" philosopher, though not in the sense of having been a disciple of Aristotle, whom he nevertheless admired greatly. Nietzsche was peripatetic in the literal sense of the word, since most of his work was not only thought but written down in small notebooks during his long walks in solitude. On such occasions he might even have thought aloud with accompanying gestures. This strong habit of his certainly helps us understand why virtually all of his writings take the form of aphorisms, some longer than others – perhaps a few pages – but most of them short paragraphs or even sentences. But there is another more deliberate reason for Nietzsche's employment of aphorisms. Since he viewed the system-building philosophies of the past as defective, he regarded his own task differently. The point was not to erect a new system – a futile enterprise in all circumstances – but to expose the flaws of the old and propose a few affirmative guidelines in the form of thought-experiments. The aphoristic "method" certainly served Nietzsche well in expressing his ideas succinctly and epigrammatically. But there is also a disadvantage to this method, as is illustrated, for example, in his aphorism on Aristotle. He salutes and honors him, but alleges that the great philosopher of antiquity had not only failed to hit the nail on the head but missed the nail itself when he indicated the ultimate end of Greek tragedy. Just look at the Greek tragic poets, says Nietzsche, and it will be easy to see what it was that most stimulated their industry, inventiveness, and competition: "certainly not the attempt to overawe the audience with sentiments. The Athenian frequented the theater *in order to hear beautiful speeches*. And beautiful speeches were what concerned Sophocles: pardon the heresy" (*The Gay Science*, 80). In thus expressing his differences with Aristotle, Nietzsche merely makes an assertion, but presents no argument, analysis, or evidence. This is fairly typical of Nietzsche's presentation of his ideas.

However, another type of criticism goes to the very heart of Nietzsche's philosophical outlook. For Nietzsche, there are moralities, but no Morality. His sociological conception of the origin of morality is strikingly similar to that of Emile Durkheim. For Nietzsche, to behave morally is to obey a certain code. Morality is custom. In *Human, All too Human* (96), for example, Nietzsche asserts that to be moral, virtuous, or ethical means to obey a long-established tradition or law. It is immaterial whether one obeys readily or reluctantly; it is sufficient that one obeys. The "good" are those who follow the customs readily, as if by nature; the "evil," those who resist custom and tradition for whatever reason. In *Assorted Opinions and Maxims* (89) Nietzsche maintains that custom is, in essence, that which is good for the *community*. In a formulation from which Durkheim would not have dissented, Nietzsche writes:

> The origin of custom may be traced to two ideas: "the community is worth more than the individual" and "a long-lasting advantage is to be preferred over a fleeting one"; from which the conclusion is drawn that the long-lasting advantage of the community must take unconditional precedence over the advantage of the individual . . .

The problem with such formulations is that they tend to justify whatever constraints a given society imposes on its individuals. This sociologistic conception of things refuses to recognize trans-historical or trans-cultural criteria by which to evaluate the customs and political practices of a given society. And in the absence of such criteria one is left with a moral relativism which renders one intellectually defenseless against, for example, the oppressive and murderous tyrannies with which we are so familiar from the history of the twentieth century. Whether Nietzsche dealt adequately with the dangers of relativism is a central question that will occupy us throughout this re-examination of his works.

It appears that by the time he completed *Die Morgenröte* (*The Dawn* or *Daybreak*) (1881) Nietzsche got no farther than trying to show that the so-called "higher qualities" of humans – those for which a transcendental origin had been traditionally claimed – were simply the transformation of "lower" qualities, those which humans have in common with the animals. It was therefore the "will to power" which appeared to offer the widest scope for human development. This suggests that by the time of *The Dawn* Nietzsche had not advanced beyond the notion that morality emerged from the desire for power and the fear of disobedience, a quasi-Hobbesian idea. All actions, says Nietzsche, may be traced back to evaluations which we have adopted.

> Why do we adopt them? From fear – that is to say, we regard it more advisable to pretend they are our own, and accustom ourselves to this pretense, so that at last it becomes our nature. (*The Dawn*, 104)

> If we ask how we became so fluent in the imitation of the feelings of others, there is no doubt about the answer: man, as the most fainthearted of all creatures due to his delicate and fragile nature, has in his *faintheartedness* the masterful teacher of that empathy, that quick grasp of the feelings of another human being (and of animals). (*The Dawn*, 142)

> Behind the fundamental principle of the current moral fashion: "moral actions are actions performed out of sympathy for others," I see the social effect of faintheartedness hiding behind an intellectual mask: human weakness and timidity desire, above all, that *all the dangers* which life once posed should be removed from it . . . (174)

> My rights . . . are that part of my power which others have not merely yielded to me, but which they want me to hold on to. How do these others arrive at that? First, through their good sense, fear, and caution: whether in that they expect something similar from us in return (protection of their own rights); or in that they recognize that a struggle with us would be perilous or to no purpose; or in that they see in any reduction of our power a disadvantage to themselves, since we would then be incapable of forming an alliance with them in opposition to a hostile third power . . . That is

how rights originate – in recognized and guaranteed degrees of power. When power-relationships undergo any significant change, rights disappear and new ones are created – as is shown in the continual disappearance and reappearance of rights among nations. . . . If our power appears to be profoundly shaken and shattered, our rights cease to exist: conversely, if we have become very much more powerful, the rights of others, that we have previously conceded them, cease to exist for us. (112)

This aphorism is little more than a reworking of Hobbes's proposition that the international arena is in a "state of nature," in a "war of each against all," for the fundamental reason that it lacks a Leviathan. Even the striving for distinction

is the striving for domination over another individual, though it be a very indirect domination . . . (113)

We see, then, that Nietzsche's view of morality, as derived from power, owes a great deal to Hobbes. There is, however, a very important difference between Hobbes and Nietzsche, a difference related to our earlier observation that the primary responsibility of a philosopher is to give good reasons for what he believes. Although Nietzsche is zealous in his determination to expunge all metaphysics from his rethinking of things, his central notion of a "will to power" is more in the nature of an ambiguous metaphysical postulate than a rationally and empirically grounded proposition. Hobbes, in contrast, not only defines power as an individual's "present means to obtain some future apparent good," he also provides reasons for his view that there exists in all individuals a natural and restless desire for power that ceases only in death. Indeed, Hobbes explains why individuals pursue more and more power: not only because they hope for a greater delight that increments of power will bring them, but also because they cannot secure the power they already have without acquiring more.

There are, then, utterances in *The Dawn* which suggest that Nietzsche had, in effect, reduced morality to power-relations. He reminds us, in that respect, of two proto-Nietzscheans with whom Socrates had to contend, Callicles and Thrasymachus, who had asserted that "might is right." We shall examine those dialogues later to see how Socrates sought to refute that assertion. But there are other statements in *The Dawn* indicating that Nietzsche held implicit standards, though he refused to call them moral. Nietzsche denies both morality and immorality. He does not deny

> that countless people *feel* themselves to be immoral, but that there is any true reason so to feel. Understandably, I do not deny – unless I am a fool – that many actions called immoral ought to be avoided and opposed, or that many actions called moral ought to be done and promoted – but I think that the former should be avoided and the latter promoted *for other reasons than hitherto*. (103)

But, alas, Nietzsche not only fails to inform us which actions should be opposed and which promoted, he also fails to tell us what his "other reasons than hitherto" are. Does he have standards or criteria for making the distinction? Or is he saying that desirable and undesirable actions are merely a matter of taste – *his* taste? And yet, there is at least one aphorism in *The Dawn* strongly suggesting that Nietzsche did in fact hold certain ideals and standards which, he believed, were "in accordance with the commandments of reason":

> There are today among the various nations of Europe perhaps ten to twenty million people who no longer "believe in God" – is it too much to demand that they should *give a sign* to one another? Once they have thus come to *recognize* one another, they will also have made themselves known to others – they will at once become a power in Europe and, happily, a power *between* the nations! Between the classes! Between rich and poor! Between rulers and subjects! Between the most unpeaceful and the most peaceful people. (96)

If we take this statement at face value, it is truly remarkable for what it reveals about Nietzsche's vision, at least at this stage of his thinking. It reveals, first of all, that peace between the nations and classes was for Nietzsche an ideal or a good that one *ought* to try to achieve. But if Nietzsche denies morality and immorality, as he claims, how would he justify the pursuit of this ideal? Where nations and rich and poor are concerned, this ideal is, after all, as old as the vision of Isaiah and the other Hebrew prophets of the eighth century BC. But if, as we shall see, Nietzsche regards much of what we find in the Jewish and Christian traditions as a "slave morality," a morality, indeed, which needs to be supplanted by the vision of Zarathustra, he can scarcely ground his ideal in those traditions. And if the actions he advocates in pursuit of his ideal are to be promoted "for other reasons than hitherto," why does he refrain from telling us what those reasons are?

But Nietzsche's statement is remarkable for another reason. That Nietzsche should have rested his hopes for the future of Europe on the millions of people who no longer "believe in God" strikes one as the height of naïveté. Presumably Nietzsche was referring to the large numbers of Europeans who attended church irregularly or not at all, who perhaps preferred civil marriage ceremonies to church weddings, who refrained from baptizing their children, and so forth. In a word, Nietzsche had in mind the secular atmosphere of much of Western Europe's urban social life. If there were, in fact, millions of such people, one cannot doubt that most of them were decent individuals who would have approved of Nietzsche's ideal of peace. But there are at least two observations which need to be made about Nietzsche's ideal, the first being that secularism, materialism, and atheism are not prerequisites for human decency. For, as Dostoevsky so powerfully demonstrated in his great novels, it is quite possible for someone who has lost his faith to believe – as did Raskolnikov in *Crime and Punishment* – that if God is dead, everything is permitted. The danger of this kind of moral nihilism Nietzsche nowhere takes into consideration. The history of twentieth-century Europe has demonstrated that the most militantly atheistic regimes were also the most tyrannical and murderous. Because Dostoevsky anticipates Nietzsche in dealing with the decline of the Christian faith, and confronts the resulting danger of moral nihilism directly, a later chapter is devoted to that issue.

In *The Gay Science* (*Die fröhliche Wissenschaft*), one finds in embryo several of Nietzsche's chief ideas: the "will to power," the "higher type" or the "superman," and the "eternal recurrence." "The strongest and most evil spirits," writes Nietzsche,

> have so far done the most to advance humanity: again and again they reignited the passions that were falling asleep – every ordered society puts the passions to sleep – and they reawakened again and again the meaning of comparison, of contradiction, of the pleasure in what is new, daring,

untried; they forced men to pit opinion against opinion, ideal against ideal. In most cases [they accomplished this] by force of arms, by destroying boundary markers, by violating pieties – but also by means of new religions and moralities. (*The Gay Science*, 4)

In this collection of aphorisms we hear once again that there are only "moralities," but no Morality as yet. As "Morality," so-called, has been deprived of its transcendental origin and sanction, it can have no everlasting and universal worth. To regain such worth it must emerge as the consequence of "necessity" felt by those who frame it. And who are the future framers? In one of his most militantly rhetorical statements Nietzsche writes:

I welcome all signs that a more manly, warlike age is about to dawn, which will restore honor to courage above all. Then that age will pave the way for one yet higher, and it will gather the strength that the higher age will find necessary – the age that will carry heroism into the pursuit of knowledge and that will *wage wars* for the sake of ideas and their consequences. To that end we now need many valiant human beings . . . who are determined to seek in all things what must be *overcome* in them. . . . For, believe me, the secret for harvesting from existence the greatest fruitfulness and enjoyment, is to *live dangerously*! (*The Gay Science*, 283)

The war metaphors which recur frequently in Nietzsche's writings have harmed his reputation. But Walter Kaufmann and R. J. Hollingdale have sought to defend him, arguing that the war rhetoric simply expresses the same idea he had held as a schoolboy, inspired by Heraclitus' doctrine that strife is a necessary component of the creative process.

And as a result of the careful and detailed research of those scholars, the following characteristics of Nietzsche's outlook and conduct are beyond doubt. He was neither a racist nor a German nationalist. On the contrary, in both his published writings and his notes he attacks antisemitism several times, and goes out of his way to praise Jews, Judaism, and the Hebrew Bible. Thanks to Kaufmann's work we now understand clearly how the blatant distortions of Nietzsche's work came about. During the years of Nietzsche's insanity and incapacity, and after his death, his sister Elizabeth became the executor of Nietzsche's Literary estate. Together with her husband, a notorious Jew-hater named Bernhard Förster, Elizabeth "edited" and tampered with Nietzsche's writings, interpreting his "will to power" and his war rhetoric in such a manner as to make of him a proto-Nazi theorist. Kaufmann and Hollingdale therefore suggest that while Nietzsche's words may have been unwisely chosen and lend themselves to distortion, his real meaning is that the "new" philosopher of the higher type must not only be a thinker but a "warrior" as well, aggressively living his philosophy and showing the way. Nevertheless, as we shall see in due course, the fact that Nietzsche's examples of "higher types" are so often individuals like Alcibiades, Julius Caesar, Cesare Borgia, and Napoleon, raises serious doubts whether he merely had philosophers in mind.

In *The Gay Science*, then, we find the earliest formulations of Nietzsche's mature ideas. These include that of "eternal recurrence," which addresses the question of how one would respond to a demonic edict. Nietzsche makes his point by asking how we would feel if one day or night a demon were to intrude upon us and dictate that the life we have so far lived we must live innumerable times more, with nothing new in it; that we must relive not only every joy but every pain, every thought, every event small or great, all in the same succession and sequence, as the eternal hourglass of existence turns over and over, and us with it. How would we respond to such an edict? Would we cast ourselves down wailing, gnashing our teeth and cursing the demon who delivered that edict? Or would we experience this event as a great moment, and respond in delight that we have never heard anything more divine? Would such an edict crush us, or bring us great joy?

It is highly probable that Nietzsche believed literally in the eternal recurrence of events throughout the universe. If he did, and if this rather crudely mechanistic notion has any philosophical significance at all, the message seems to be this: one should strive to live one's life in a manner so as to welcome such an edict. If, however, "eternal recurrence" is inexorably cyclical as Nietzsche apparently believes, does this concept contradict other key Nietzschean ideas? We shall need to return to this question since the notion of the "eternal recurrence" recurs frequently in Nietzsche's later writings.

NOTES

1 See Walter Kaufmann, *Nietzsche: Philosopher, Psychologist, Antichrist*, 4th edn (Princeton University Press, Princeton, N. J., 1974); and R. J. Hollingdale, *Nietzsche: The Man and His Philosophy* (Louisiana State University Press, Baton Rouge, 1965). I rely on Hollingdale's splendid study for these details about Nietzsche's childhood and his intellectual development.
2 Hollingdale, ibid., p. 31.
3 Ibid., p. 77.
4 Throughout, numbers in parentheses refer to aphorisms, not pages.

2

Thus Spoke Zarathustra

It was during two of his customary and most enjoyable walks, Nietzsche informs us, that the first part of *Zarathustra* had occurred to him in its entirety, and especially the figure of Zarathustra as a type. To give us some insight into the type, Nietzsche begins by underscoring that Zarathustra possesses *great health*. Nietzsche's new goal for humanity requires, as a new means, a new health that is stronger, tougher, more audacious, and also more joyful than any previous health. Present-day man is simply not up to the task of charting the uncertain future. What is needed is a new spirit who, out of abundant and overflowing power, is willing to play naïvely with everything that has hitherto been called holy, good, and divine, a spirit who is audacious enough to see danger, decay, and debasement in the value standards people have naturally accepted till now. Hence, this new and higher type envisioned by Nietzsche is characterized by a superhuman well-being and benevolence – which might, however, occasionally appear inhuman – and a great seriousness in confronting and transcending present-day moralities.

Zarathustra, as a higher type, sees farther and possesses greater capabilities than any other human being. In him humanity has been overcome, and the concept of "superman" has become so superlative a reality that whatever has so far been considered great in humanity lies beneath the superman at an infinite distance. What is "man", then, for this Superman? Neither an object of love nor, worse, of pity; for Zarathustra has mastered the great "nausea" over humanity which for him is simply an "unform, a material, an ugly stone that needs a sculptor" (*Ecce Homo*, "Thus Spoke Zarathustra," 8). For Nietzsche, then, Zarathustra is confronted by a *Dionysian* task, and the indispensable condition for such a task is "the hardness of the hammer, *the joy even in destroying*. The imperative, 'become hard!', the most fundamental certainty *that all creators are hard*, is the distinctive feature of a Dionysian nature" (ibid).

From this paraphrase of Nietzsche's own view of Zarathustra as a new and ideal human type, we can see that he has, in a sense, returned to his earliest intellectual preoccupations in *The Birth of Tragedy*. There he had argued that the development of the Greek genius was determined by the Apollinian-Dionysian duality. This was especially true of Greek tragedy. Then Socrates entered the picture, and the older dualism was replaced by a new one: the Dionysian-Socratic. Socrates' aesthetic principle maintained that "to be beautiful a thing must be intelligible." Art is no exception to the principle that "knowledge is virtue." It was under the influence of this doctrine, Nietzsche argues, that Socrates' contemporary Euripides began to measure all the separate elements of drama – language, character, structure, choric music – and organize them according to the Socratic principle. It was this "audacious reason-ableness" that led, after Sophocles, to the conspicuous degeneration of Greek tragedy. Dionysian passion had been crushed by Socratic reason, and the great, creative art of the Greek drama was wrecked as a consequence. There is a powerful tendency in Nietzsche to think in dualisms, which suggests why he chose "Zarathustra" as his super-protagonist.

Among the great founders of the world religions, it was Zarathustra (Zoroaster) who founded a dualistic religion based on the strife between the antagonistic principles of light and darkness, Good and Evil, Ormazd and Ahriman.[1] The little that we know about ancient Iran in the first millennium BC suggests that the people whom the prophet addressed with his new teachings were in transition from a nomadic or semi-nomadic state of existence to that of tillers of the soil. The prophet's primary mission, it appears – though certainly out of internal, spiritual promptings – was to strengthen the new economic and social trend. The older idolatry that was bound up with the nomadic life was prohibited by the prophet in absolute terms. There is a great god and a master of the world, he taught, who is the creator and provider of all life on earth. He is the good god, the god of light, Ormazd. Anyone and anything that opposes the light and the good participates in the work of Ahriman, the god of evil and darkness and the leader of the demonic forces which, according to the historical Zarathustra, were the deities of the old Indo-Persian faith, the religion of the Magis, which continued to exist in Iran side by side with Zarathustra's new teachings. In Zarathustra's new doctrine, then, there are two domains at war with one another. The domain of Ormazd is that of the true and just order of the cosmos. The ancient Greeks interpreted Zarathustra's teachings to mean that Ormazd was responsible for *eunoia*, a term comprising both justice and kindness, and also responsible for *eunomia*, the good order of the universe. The domain of Ahriman, in contrast, is that of the *lie*, and his power takes the personified form of a demonic figure. Herodotus had observed in this regard that the Persians "consider telling lies more disgraceful than anything else . . ." (*The Histories*, I, 138).

Now we come to an important point in the historical Zarathustra's doctrine that many be pertinent to Nietzsche's adaptation. In the momentous struggle between the two gods and their domains, human beings are obliged to stand on the side of the god of light and goodness. When they sow and plant and build and bring fertility to the desert and wilderness, they thereby strengthen the powers of Ormazd over Ahriman. The more they engage in such constructive activities, the more they increase the power of good over evil, light over darkness, and the sacred over the profane. For Zarathustra is reputed to have said: "the sower of grain sows holiness." The tillers of the soil, the farmers and the herdsmen, are the workers of light who expand the good in this world; the tent-dwelling nomads, who do nothing to fructify the earth but, on the contrary, often plunder the fruit of the peasants' toil, are those who strengthen the forces of darkness and evil. Thus Zarathustra taught that one ought to treat with affection the useful, non-predatory, domesticated animals, and especially the ox that pulls the plough and, of course, the quiet and peaceful cow that nourishes the human being with her milk. To the extent that human beings are productive and constructive, the spirit of Ormazd is victorious and that of Ahriman is vanquished. The final victory, however, of Good over Evil will be achieved by Ormazd himself, thus creating the kingdom of one supreme god over the world. In this final victory of Ormazd he will be assisted by three "messianic" figures who are the "sons of Zarathustra." The prophet, Zarathustra, thus laid great stress on the positive virtues of justice and kindness, and he appears to have had considerable influence on the culture of his time. Furthermore, his doctrine tends in the end to transcend the original dualism in favor of the total victory of goodness and light.

As we now turn to Nietzsche's protagonist, we need to ask whether Nietzsche, too, abandoned and repudiated his original dualism. For, as we have seen, in *The Birth of Tragedy* Nietzsche had argued that the older Apollinian-Dionysian dualism was supplanted by the Dionysian-Socratic opposition in which Dionysian passion was overwhelmed by Socratic reason, with the consequence that Greek drama was destroyed. Clearly, Nietzsche does not look with favor upon this development, since he regards Dionysian passion as the most fundamental element of the creative process. And as we have seen, in *Ecce Homo* – one of his final reflections on the nature of his protagonist – Zarathustra was faced with a Dionysian task which required, above all, a *Dionysian nature*. This suggests that, like the historical Zarathustra, Nietzsche abandoned his earlier dualism, but that in his case Dionysus defeats both Apollo and Socrates: the demon of darkness overpowers the forces of the sun-god, and reason is no longer recognized as the supreme principle and standard of values. Accordingly, as we shall see in our re-examination of Nietzsche's *Zarathustra*, he now temporarily abandons his aphoristic method which, despite its shortcomings, was characterized by an "Apollinian" measured thoughtfulness, and replaces it with self-styled Dionysian dithyrambs.

In *Thus Spoke Zarathustra*, Nietzsche wishes to accomplish both a negative and a positive task. He wishes to topple the twin pillars of Western ethics, the Christian and the Platonic legacies, and he wishes to put in their place a new and stronger philosophical foundation for life. What is the nature of Nietzsche's positive message? In the Prologue he speaks again of the now familiar death of God, and he proclaims the superman. If faith is lost, as Nietzsche repeatedly claims, then it is up to human beings themselves to give their lives meaning. How? By raising themselves above the animals and the all-too-human. But what is the new and superior meaning that Nietzsche now offers us? That meaning, such as it is, lies in the few human beings who raise themselves above the all-too-human mass.

PART ONE

In Part One, Zarathustra declares, "I love man." And the saint, his foil, says no, man is too imperfect, so I love God. Zarathustra responds by explaining that his love consists in bringing men a gift, though the nature of the gift remains temporarily unspecified. The saint, as a representative of the "decadent" tradition which Nietzsche wishes to discredit, reacts as one might expect: what men need is not a gift, but help in shouldering their burdens; that is why he composes and sings songs of praise to God. And Zarathustra is puzzled; is it possible that the saint has not yet heard that God is dead?

In the next town a crowd had gathered in the marketplace waiting to be entertained by a tightrope walker, so Zarathustra, the prophet, takes advantage of the opportunity to address the people thus: I teach you the superman! Just as the ape is a laughing stock to man, so shall man be for the superman. Indeed, the superman *is* the meaning of the earth and *shall be* its meaning. Zarathustra beseeches his listeners to be faithful to the *earth* and to pay no attention to those who speak to them of otherworldly hopes. To sin against the earth is now the most dreadful thing. "What is the greatest experience one can have?" Zarathustra asks. This is the time, he explains, when one should acquire the attitude of contempt and disgust for the traditional conceptions of happiness, reason, and virtue. What do these matter? Zarathustra urges his audience to throw off their concerns about good and evil and sin, and to turn their faces toward the sky so as to be licked by the tongue of lightning and inoculated with frenzy. For I teach you the superman, the prophet declares; he is this lightning, he is this frenzy. Man is a rope, tied between the beast and the superman; man is a bridge to that higher specimen. Whom does Zarathustra love? Only those who live to know that the superman may live some day. Uncertain whether the people understand him, the prophet speaks to them of what is most contemptible – the *last man* – who is most despicable because he cannot despise himself.

Faced with a discouraging lack of understanding on the part of the people, Zarathustra has a sudden insight: he should speak not to the people, but to companions. Zarathustra will not become the shepherd and dog of a herd. No, his mission is to lure the few away from the herd.

The herd will be angry with him as will be the shepherds of the old "true faith." But that is what the prophet wants; anger in response to his message is only to be expected, after all. Whom do the believers of all faiths hate most? The individual, of course, who breaks their table of values. Zarathustra, who is himself a creator, now seeks fellow creators who will write new values on new tablets. Zarathustra will no longer waste his time and energy speaking to the people, which is like speaking to the dead. Instead, he will seek creators and show them the steps to the superman.

In one of his several speeches, Zarathustra then speaks of the "afterworldly." He acknowledges that at one time he too had cast his mind beyond man. It was an illusion, a form of madness like all after-worlds and god-beliefs. Worse, it was a reflection of a poor specimen of man. But in time, Zarathustra proclaims, he overcame himself, recognizing that the ghost he created had not come to him from beyond. It was then that he ascended the mountains and created a brighter flame for himself and, behold, the ghost disappeared. Others, too, must come to understand, as did Zarathustra, that all afterworlds are created by suffering and helplessness. Not only suffering but weariness too – an ignorant weariness that wants to attain an ultimate condition with one leap. That is what has given birth to all gods and afterworlds. Ignorant weary souls have failed to realize that the otherworld they longed for was a dehumanized, inhuman world, a heavenly nothingness. Just as Zarathustra's ego has acquired a new strength and pride, others too must learn no longer to bury their heads in the sands of heavenly nothing, but to carry proudly an earthly head capable of creating meaning for the earth. It was weakness, sickness, and decay which had led to contempt for the body and for the earth, and which invented a heavenly realm and redemptive drops of blood. Zarathustra therefore implores his audience to listen not to the sickly but to the voice of the healthy body, which speaks most honestly and purely when it speaks of the meaning of the earth.

The ignorant and innocent speak of body and soul; but those who have awakened from their delusive dreams know that they are body and nothing else. Zarathustra assures his listeners that the body is a mighty ruler who stands behind their thoughts and feelings. Make no mistake, there is more reason in one's body than in one's best wisdom. Who, then, are the despisers of the body? Only those who want to die and who turn away from life because they are incapable of creating beyond themselves, incapable of self-overcoming. These weary and weak specimens are certainly no bridge to the superman.

Unlike the despisers of the body, who call the passions evil, those who recognize the wisdom of the body love the passions. It is the body's wisdom that tells us to laugh and to feel elevated. Zarathustra would believe only in a god who could laugh and dance and be thus elevated in contrast to a devil whose solemnity and gravity make all things fall. Our passions, including our so-called wicked instincts, thirst for freedom. But there are preachers of death who preach against life because they carry within themselves those passions and instincts – those beasts of prey – and who are torn between their lust and their self-laceration. They fear their own bodies and, therefore, have not as yet become real human beings. Then there are those whose inner self is so weak that they believe that life is refuted by the presence of the sick or the old. They live in deep melancholy, and appear almost eager for an accidental event that will bring death. Still others see life as suffering and nothing more.

Zarathustra wants to teach those who will listen and understand, that there is hatred and envy in their hearts, and that they must be big enough to recognize that fact without shame. And the prophet now speaks to them of war and warriors: "Love peace as a means to new wars, and the short peace more than the long." He recommends not work but struggle, not peace but victory. "You say it is the good cause that hallows even war? I say unto you: it is the good war that hallows any cause. War and courage have achieved more great things than love of the neighbor." These are some of the utterances that have harmed Nietzsche's reputation. But there is no honest way in which his real meaning can be twisted here, for he starts the whole passage by saying: "Your enemy you shall seek, your war you shall wage – *for your thoughts*" (Part One, "On War and Warriors").

Zarathustra now calls his listeners' attention to another danger threatening his vision: the new idol, the nation-state, which he calls the coldest of cold monsters, and a liar too, for it claims that the state is the people. Where the state ends, that is where we will see the rainbow and the bridge to the superman. But, alas, in his travels among many peoples the prophet observed a thousand and one goals: what was good to one people was scorn and infamy to another. A thousand goals for a thousand peoples, but the *one* goal is lacking. "Humanity still has no goal" (I, "On the Thousand and One Goals"). So Zarathustra can speak of "humanity" and the need for a universal goal; but he can also say: "The delight in the herd is more ancient than the delight in the ego; and as long as the good conscience is identified with the herd, only the bad conscience says: I" (ibid). Zarathustra has by now acquired disciples at whom he looks lovingly, urging them once again to allow their gift-giving love and knowledge to serve the meaning of the earth. He now calls for a new "chosen people" out of whom the superman may emerge.

PART TWO

Throughout, Zarathustra speaks to his audience in what might be termed heightened prose – not quite poetry, although that is what his creator appears to be striving for. Nietzsche is recognized as one of the great prose stylists in the German language. He himself said of this work that "perhaps the whole of *Zarathustra* may be reckoned as music . . ." (*Ecce Homo*, 'Zarathustra,' 1). The result is an extraordinary range of metaphors and images drawn from the phenomena of nature and the human condition.

In Part Two, as earlier, Zarathustra tells his listeners that we used to say "God" when we beheld the awesome spectacles of nature, but now we must say: superman! "God" is a mere conjecture, and our conjectures must not reach beyond our creative will. Are we capable of creating a god? Clearly we are not and therefore ought not to speak of God or gods. Ah, but we are capable, Zarathustra assures us, of creating the superman. Well, perhaps we ourselves are not directly capable of that, but we are indirectly capable as forefathers. As precursors of the superman we must begin to expel from our minds and bodies the older thoughts and feelings that originated with the God-conjecture – pity, for example. Those who are merciful and who feel blessed in their pity are lacking in shame. Zarathustra, being a specimen of "great health," prefers the company of those who do not suffer; and although in the past he may have done this or that for sufferers, he derived no joy from his actions. From the beginning, human beings have felt too little joy. That is the true original sin. Zarathustra reminds his friends that when one sees a sufferer suffer, one is ashamed on account of his shame, and when one comes to his aid, one injures his pride. Indebtedness to those who help us evokes not gratitude but vengefulness; even the small charity one receives turns into a gnawing worm. Therefore, Zarathustra teaches, be a giver of gifts as a friend to a friend, but reserved in giving and accepting charity. And if you have a suffering friend, be a hard not a soft bed for him, and thus you will profit him best.

Everything associated with the old code of values should be abolished – especially sin, bad conscience and other doctrines of the priests who are bound in the fetters of false values and who teach "sinners" to crawl up the stairs of the church on their knees. The priests know not how to teach the love of their god except by crucifying man. It is these sickly doctrines which have prevented the rise of a superman. Zarathustra has seen both the greatest and the smallest man in their nakedness: they are all-too-similar to each other. Even the greatest Zarathustra found to be all-too-human. Under the influence of these inauthentic doctrines, those who strive for virtue want to be paid. They want heaven for earth and the eternal for their today. Zarathustra teaches that there is no reward, and he repudiates the Socratic doctrine that virtue is its own reward. Alas, reward and punishment have been laid into the foundation of things and into one's soul. One should love virtue as a mother loves her child; but what mother has ever wished to be paid for her love?

If Zarathustra teaches the superman and the "higher type," he also teaches contempt for what he calls the "rabble." All wells from which the rabble also drink are poisoned. The knowledge that life requires hostility, torture, and death causes Zarathustra less nausea than the knowledge that life requires the rabble. And now we hear Zarathustra (and, of course, his creator) letting us know what he thinks of the rise into politics of the *demos*. What is nowadays called "ruling," the prophet remarks with repugnance, is higgling and haggling for power with the rabble. More repugnant to the prophet is the discovery that even the rabble has *espirit*. There is the power-rabble, the writing-rabble, and the pleasure-rabble, all of whom Zarathustra despises and shuns. He rids himself of his nausea by climbing to heights where no rabble sits by the well, the highest spheres where the rabble cannot drink. Those who preach "equality" are actually and secretly vengeful. They wish to wreak vengeance on those whose equals they are not. Those preachers call themselves good and just, but one must not forget for a moment that they would be tyrants if only they had the power. Human beings are not equal – that is the true meaning of justice. Nor will they ever become equal.

Here, we must pause for a moment and interrupt the flow of Zarathustra's discourses. In the preceding chapter we remarked that Nietzsche's "will to power" appears to be more in the nature of a meta-physical postulate than a well-grounded rational or empirical proposition, since Nietzsche, unlike Hobbes, never explains why human beings may want more and more increments of power. Here, in section seven of Part Two, called "On the Tarantulas," Nietzsche writes: "That struggle and inequality are present even in beauty, and also war for power and more power: that is what he [the wisest man] teaches us here in the plainest parable."

What is especially interesting about this quasi-Hobbesian formulation is that it flatly contradicts Hobbes where the question of equality and inequality is concerned. Hobbes insists that the natural condition of humanity is one of *equality*. In the pre-social state of nature which he postulates, humans are equal in the faculties of both body and mind. True, one man may be physically stronger or mentally quicker than another.

> Yet when all is reckoned together, the difference between man and man, is not so considerable, as that one man can thereupon claim to himselfe any benefit, to which another may not pretend, as well as he. For as to strength of body, the weakest has strength enough to kill the strongest, either by secret machination, or by confederacy with others, that are in the same danger with himselfe.

And in faculties of mind, Hobbes saw even greater equality among humans:

> For Prudence is but Experience; which equal time, equally bestows on all men, in those things they equally apply themselves unto. That which may perhaps make such equality incredible, is but a vain conceit of one's own wisdom, which almost all men think they have in a greater degree, than the Vulgar . . . (*Leviathan*, ch. 13)

Did Nietzsche share this "vain conceit"? If all he meant by his assertion of human inequality is that not every individual can become a

Hobbes or even a Nietzsche, one can scarcely disagree, and surely Hobbes would have acknowledged genius. But would Nietzsche, who speaks contemptuously of the "rabble" and the "herd," have acknowledged what the conservative thinker Hobbes had to say about the human faculties of body and mind? The evidence throughout suggests that Nietzsche's longing for the superman prevented him from ever considering the Hobbesian view objectively. Nietzsche simply clings to his dogma that humans are not equal and that what the "higher types" may claim to themselves others have no right to claim.

Zarathustra teaches that the "will to power" is a matter of *self-overcoming*. Even those things we esteem more than life itself are manifestations of the will to power. The prophet continues to din into his listeners' ears that good and evil are mere words and that standards of good and evil that are not transitory do not exist. Zarathustra laughs at the weaklings who think themselves good because they have no claws, and he demands the good precisely from those who are strong and capable of all evil. What Nietzsche's protagonist calls "evil" are the biologically-rooted, powerful passions that Freud later called the "id" and the "libido." Indoctrinated with the old values as they are, those who pretend to love the earth and the earthly are actually consumed by shame and bad conscience. The old spirit has persuaded them to despise the earthly, but their passions, which are stronger, have not been persuaded. The prophet therefore calls them hypocrites and lechers who slander desire because they lack innocence in their desire. They dare not believe in themselves and in their "entrails," so they pretend purity, hiding from their nature behind a god's mask.

PART THREE

The earlier prophets, Zarathustra declares, taught softness and other debilitating values, promising a land flowing with milk and honey. That accounts for man's weakness, for the more he spares himself, the more sickly he becomes. Zarathustra therefore teaches hardness, which is necessary for every climber of mountains who wants to ascend to the highest peaks where he will see clearly how to annihilate the old table of values. One cannot be a follower of Zarathustra and yet wish to live without danger. One must be bold and undertake even life-risking ven-

tures, whether as searchers or researchers. The prophets and philosophers of softness have taught for thousands of years that all human
beings stand on the edge of an abyss into which they may be plunged at
any moment by myriad unexpected and irresistible misfortunes. That is
why humans have needed to believe in a power greater than themselves,
to whom they could cry out for help. But Zarathustra repudiates this old
belief as a belief in nothingness. He agrees that humans can nowhere
avoid standing at the edge of an abyss, but his consolation is hard:
courage slays dizziness and fear at the edge of abysses. The *soft* prophets
and philosophers who preceded Zarathustra also affirmed life. They
too recognized that in all times and places most human beings valued
life over death, health over sickness, happiness and joy over pain and
suffering. But they also recognized that life, health, and happiness are
unavoidably bound up with their negations, which are essential facts of
existence. William James, exploring the source of the religious experience, wrote in this regard that the

> fact that we *can* die, that we *can* be ill at all, is what perplexes us; the fact
> that we now for a moment live and are well is irrelevant to that perplexity.
> We need a life not correlated with death, a health not liable to illness, a
> kind of good that will not perish, a good in fact that flies beyond the goods
> of nature.

Hence, James observed, the essence of the religious problem is: "Help!
Help!"[2] The crying need for help is met by the belief in immortality and
by the belief in benevolent spirits and deities who can overcome the evils
besetting human existence. In his awareness of his limited, corporeal,
earthly existence, the human individual is unique in the animal kingdom.
By means of religious beliefs he gains a sense of safety and peace. In his
religious attitude, the individual views the visible world as a part of a
greater spiritual universe, the real source of his life's meaning, and his
hope for salvation.

But of course Nietzsche's Zarathustra finds all that illusory and
nauseating and will have none of it. Courage, for Zarathustra, slays not
only fear and dizziness at the edge of all the abysses of life; courage "slays
even death itself, for it says "was *that* life? Well then! Once more!' " (Part
Three, 1, "On the Vision and the Riddle"). Zarathustra thus teaches *no*
fear of death, for he is the prophet of the "*eternal recurrence.*" Some

Nietzsche scholars have dismissed this notion as a silliness that has nothing to do with the main current of Nietzsche's thought. Others have argued that Nietzsche's description of the recurrence, wherever it occurs in his writings, is a rather mechanistic formulation of the valid insight that each and every moment, as it is lived, has an infinite value. But if one carefully examines the several contexts in which Nietzsche either describes or alludes to the "recurrence," it seems indisputable that he believed literally that every moment is repeated eternally and, there- fore, that every moment is an eternity. "Behold this moment," says Zarathustra,

> From this gateway, Moment, a long, eternal passage-way leads *backward*: behind us lies an eternity. Must not whatever *can* walk have walked on this path before? Must not whatever *can* happen have happened, have been done and passed by before? And if everything has occurred before, what do you think of this moment? Must not this gateway also have been there before? And are not all things so tightly tied together that this moment draws after it *everything* that is to come? Therefore – itself too? For whatever *can* walk in this long passage-way out *there* too, *must* walk once more. (Part Three, "On the Vision of the Riddle," 2)

It is precisely because of Nietzsche's belief in the reality of the "eternal reccurence" that he can say that all things baptized in the well of eternity are "beyond good and evil." For the corollary of the "recurrence" is the lack of a principle of direction, *chaos*. Zarathustra therefore teaches his followers that they must deliver themselves from the bondage of purpose and recognize that all things stand under the "heaven" of Accident and Chance. He teaches that "freedom" is thus placed over all things and that over them no "eternal will" wills. In opposition to Purpose and Eternal Will, Zarathustra teaches the anti-Socratic doctrine that in " . . . everything one thing is impossible: rationality." There is no eternal web of reason. What all this means, for the prophet, is that in the face of such "freedom" the human being must be big, strong, and courageous.

But instead, Zarathustra declares, everything has become smaller so that individuals of his kind must now stoop to enter through the lower gates. It is the old virtues that make everything, including the people, small. He finds it hard to accept that small people are now needed.

Indeed, they become smaller and smaller because they want contentment. The small people have become a stumbling block, for although some of them will, most of them are only dictated to. The old virtues have led to that blatant hypocrisy in which even those who command pretend that they serve. Zarathustra deplores what he sees all around him among the small people: so much kindness, so much justice, so much pity – so much weakness. What these weak souls really want is to please and satisfy everyone so that nobody will hurt them. They call this virtue but actually it is cowardice. They lack fists, these small people; they have become tame, cowardly, domestic animals. And when the prophet curses all those who are cowardly and who whine while folding their hands in prayer, they shout in defiance "Zarathustra is godless." To which the prophet proudly replies "Yes, I *am* Zarathustra the godless!" And he instructs his listeners to reject the teachers of resignation and to become the prophet's equal by giving themselves their own will. The small people, becoming smaller and smaller, may yet perish of their small virtues, abstentions, and resignations. They have yet to learn that in order for a tree to become great, it must strike hard roots around hard rocks.

Zarathustra now teaches a new lesson: there are at least three natural urges in the human individual which have been maligned by the old teachers and called evil. The three so-called evils are *sex, the lust to rule*, and *selfishness*. It is that age-old servility to the god-idea that has given birth to and nurtured an ascetic attitude towards the human body and its passions. If one wishes to overcome such sickliness, one must liberate oneself from the old virtues and recognize that these "evils" emanate from a strong, healthy, supple, triumphant, and persuasive body. The deceitful and hypocritical "love thy neighbor!" must be replaced with "love thyself!" for otherwise man is like a camel who allows such a grave, heavy, and alien burden of self-denying values to be loaded upon him that life seems a desert to him.

Man must cease his quest for a good for *all* and an evil for *all*. He must discover the true self with its lusts and passions and gain the courage to declare: "this is *my* good and evil." That is what the prophet teaches, for it was by many paths that he reached *his* truth. As he climbed and ascended ever higher he only reluctantly asked the way, and this always offended his taste. He preferred to try out the various paths by himself.

One must emulate the prophet by trying and questioning and also by answering for oneself. That is his *taste*, says Zarathustra – not a matter of good or bad, but simply a matter of taste. That is *his* way and his listeners must find theirs. To those who asked the prophet for "the way," Zarathustra replied, *the* way does not exist. He assures them that it is nothing but an old conceit among humans that they have long known what is good and evil. What is good and evil no one knows yet, unless it be the individual who creates. There is no Eternal Will, no Purpose; the universe is Accident and Chance. It is only human beings themselves, therefore, who can create purposes and goals and thereby give the earth its meaning and future. If they thus follow the teachings of Zarathustra, they will form the bridge to the superman.

We thus see a central motif in these discourses: a Dionysian emphasis on the vital passions, and a corresponding repudiation of Socratic rationality. If the universe is Chaos and Accident and there is no eternal web of reason, then rationality is a useless fiction, a non-virtue. There is no Good and Evil for *all*; only mine, yours and others'. No one Way, only many ways. In rejecting Socratic reason, Zarathustra also combats the Socratic-Platonic theory of the Forms, which was at least in part an attempt to reply to Heraclitus' dictum that one can never step into the same river twice. Constantly flowing, as it does, it is not the same river but a different one at every moment. The entire universe, said Heraclitus, is like a river, in a constant state of flux. This conception of things tended to unnerve Plato, for it denied the possibility of absolute, transcendent, lasting standards of the good, the just, the beautiful, or whatever. In Plato's dialogue *Euthyphro*, a priest by that name tells Socrates that the gods love the good, to which Socrates replies: do the gods love the good because it is good? Or is the good good because the gods love it? As the dialogue develops it becomes altogether evident that Socrates believed the former to be true – the gods love the good because it is good. The Good and the other Forms are higher than the gods; they are real, self-subsistent entities existing in the super-celestial spheres. They are the perfect transcendental standards by which the objects of sensory experience in the world may be evaluated. Obviously, Zarathustra's creator regards this theory as hopelessly metaphysical and unacceptable.

Zarathustra therefore defends the Heraclitean view by lampooning the Platonic theory. Spanning the river are bridges consisting of planks and railings. So blockheads ask, how can you say everything is in flux when the planks and railings are *over* the river? Whatever is *over* the river is stable, solid and firm. The bridges – i. e., all the values of things, the concepts of "good" and "evil" – are *firm*. And when the hard winter comes and the river freezes over, even the more intelligent tend to say "everything stands still." That is truly a winter doctrine, says Zarathustra, against which the thawing wind preaches. Like a raging bull it breaks the ice with its wrathful horns. Besides, let us not forget that before the thaw, ice breaks bridges! So everything is in flux, and all the bridges and railings have fallen into the water and flowed away. Goodbye to "good" and "evil."

Zarathustra now turns his attention to the oldest teachings that are still called holy: "Thou shalt not kill! Thou shalt not rob!" It is these so-called holy words, says the prophet, that are the real killers and robbers, for they hide the truth that all life itself is killing and robbing. In order to break the old tablets, one must resist both the despot and the rabble. The prophet therefore calls for a new nobility that will oppose all that is despotic and all that comes from the herd. The new nobility must write new noble words on new tablets. A noble view means that one must know against whom to wield the sword; it means that one shall have only enemies who are to be hated, not despised. The noble is proud of his enemies, and saves himself for the worthier enemy. The prophet thus teaches those who aspire to join the new nobility to bypass any battle concerning the rabble who raise such a clamor about the "people." The new nobility should also lay its sword to rest where the Fatherland is concerned; for it is not in the Fatherland but in the *children's land* where hope for the rise of the superman resides. Zarathustra again strikes a loud Dionysian note as he reminds the aspirants to the new nobility that human beings need what is most "evil" in them (i.e., the instincts and the passions) in order to bring out what is best in them. The evil is the power with which the highest creators are able to crush the hardest stone.

The prophet knows that he must brace himself for his great destiny, which has never yet been an individual's destiny – to teach the *eternal recurrence*, that all things recur eternally and have already existed an eternal number of times. The hourglass turns ceaselessly over and over and over again. The soul is no less mortal than the body, and the web of causes, in which every individual is enmeshed, will recur and create him or her again. Like everyone else, the prophet, too, belongs to the causes of the eternal recurrence and he comes again *not* to a new life, better life or similar life, but to this self-same life, to proclaim the superman.

But Zarathustra's audience, myself included, finds this doctrine difficult to understand and perplexing. If the universe is Accident and Chance, we ask, and if all of us will come again to this self-same life, what is the point of proclaiming the superman? It follows from the prophet's doctrine that the supermen of the past – whether Shakespeare, Beethoven, Caesar, or Napoleon – were the product of accident, not proclamation. Will they simply recur? And if there is no such thing as a new life – never mind a better life – doesn't that imply that no *new* supermen will appear? These are the crucial questions to which Zarathustra's audience demanded a convincing reply. It remains to be seen whether Nietzsche's protagonist or Nietzsche himself will provide such a reply.

In an earlier comment on the aphoristic style employed by Nietzsche in his pre-*Zarathustra* writings, we criticized that style as one which too often seemed to present us with a collection of unsupported assertions. But the style or "method" Nietzsche adopted in *Zarathustra* became problematic in his own eyes, for as Walter Kaufmann has observed: "Although Zarathustra's buffooneries [in Part Four] are certainly intended as such by the author, the thought that he might be 'only' a fool, 'only' a poet, 'climbing around on mendacious word bridges,' made Nietzsche more than despondent. Soon it led him to abandon further attempts to ride on parables in favor of some of the most supple prose in German literature."[3]

PART FOUR

The soothsayer of Part Two reappears and asserts to Zarathustra that nothing changes, nothing is worthwhile, the world is meaningless and, hence, knowledge is useless. So Zarathustra leaves in search of the "higher specimens" of man. Zarathustra continues to express his contempt for the "mob," which has now become a mob-hodgepodge in which all the social elements are mixed together: Junker, bourgeois, peasant, every species of beast out of Noah's ark. Still, the prophet has a few good words to say about the peasants of his time who embody some of those "hard" and "noble" qualities he attributes to the higher type. The peasant is best today, he says, for he is healthy, coarse, cunning, stubborn, and enduring. In the encounter with the old magician, however, the prophet informs him that he has not as yet seen a great human being, for today's realm is that of the mob, in which the people say, "Behold, a great man!" – but he in due course is exposed as a big bag of wind who, like a frog that has puffed itself up too long, will soon burst. How, asks the prophet, can one distinguish the great from the small in an era such as the present one that belongs to the mob?

Departing from the magician, Zarathustra soon meets a tall man in black with a gaunt, pale face, whom the prophet recognizes as another kind of black artist, a priest who thinks of himself as a miracle worker by the grace of God, but who is actually a world-slanderer. When the man in black tells the prophet that he has just emerged from the forest and has not yet heard the latest news, Zarathustra replies with the familiar refrain: the old God, in whom all the world had once believed, is dead! The man in black then reveals that he is the last pope: he has abandoned his post and run away in order to seek the most pious among those who do *not* believe in God – namely Zarathustra. The prophet describes the old God as a bungler who has wreaked revenge on his creations after having bungled them himself. What a sin against good taste! Impressed, the last pope acknowledges that it must be piety itself that no longer allows the prophet to believe in such a God. The prophet's great honesty, he further acknowledges, will surely lead him beyond good and evil too.

In another encounter, the "ugliest man" expresses his gratitude to the prophet for not pitying him. He honors Zarathustra for being the only prophet who teaches that pity, whether it be God's or man's, is obtrusively offensive. An unwillingness to help can be nobler than jumping to help. In anticipation of the more scathing things he will have to say in *The Anti-Christ*, Zarathustra-Nietzsche chides that preacher from Galilee, "that queer saint and advocate of the little people," who immodestly said of himself, "I am the truth," thus giving the little people swelled heads. When Zarathustra had left the "ugliest man," he felt lonely and cold, but soon found comfort among some unknown companions whose warm breath touched his soul. But when he looked around to see who these comforters were, he discovered that they were a herd of cows intently listening to a speaker. Hearing a human voice, the prophet jumped up and pushed the animals apart for fear that somebody had been injured in their midst. He was mistaken, however, for behold, there sat a man on the ground, a peaceful man and sermonizer on the mount, preaching goodness. Only if we turn back and become as cows, he was saying, shall we enter the kingdom of Heaven. Those to whom the preacher had first turned did not accept him, so he went to the cows, whom the preacher-beggar loves much but not as much as he loves Zarathustra, whom he calls good, and even better than a cow. Not flattered, the prophet brandishes his stick and drives the affectionate beggar away. This, of course, is an example of the buffoonery that Nietzsche himself eventually recognized as self-defeating. Here and elsewhere the defects of the style and "method" of *Zarathustra* become unmistakably evident: the absence of a sustained analysis or argument in support of his assertions; a "story-plot" that is not, in and of itself, very interesting; and, finally, an attempt at pervasive humor which, more often than not, falls flat. It is especially disappointing that even when Nietzsche raises an issue which undoubtedly deserves at least a paragraph or two of intelligent discussion, he gives us something else instead. Take, for example, the issue he so briefly raises in "The Shadow" (Part Four, 9). If God is dead and nothing is either good or evil or true or false, is everything permitted? Does Nietzsche have anything at all enlightening to say where this question is concerned? No. Instead of coming to grips with this question, however briefly, he tells us that he plunged into the coldest waters with head and heart, standing there afterward naked as a red crab.

As Nietzsche's narrative resumes, those on whom Zarathustra looks as the men of great longing, nausea, and disgust, come to him saying they do not want to live unless they can learn to hope again. They have come to learn from him the *great* hope. They consider themselves "higher" men who possess the last vestige of God. But Zarathustra informs them, respectfully, that they have come to the wrong prophet. They may indeed all be higher men, but they are not high and strong enough. That they need and seek hope shows that they stand on sick and weak legs. They have not learned what the prophet has worked so hard to teach: your legs and arms – your entire body – are your warriors, to whom you must show no consideration. For otherwise they become unfit for war, collapsing as soon as they hear the loud roll of the war-drums. It is not for men such as these that the prophet has been waiting in the mountains. They still carry the burden of the old doctrine. Even when they listen carefully, they distort the prophet's message, thinking that however difficult it is to understand, it must be a new philosophy of hope. So crooked and misshapen are these men, due to the remnant of God-belief in them, that they can never be hammered right and straight to the prophet's taste. It is others for whom he waits in the mountains – those who do not prostrate themselves but who stand straight in body and soul, cheerfully and triumphantly, like laughing lions. This is a fairly typical example of how Nietzsche's "method" in *Zarathustra* enables him to avoid dealing with an important question – whether a philosophy of hope is a fundamental psychological need for human beings. There are, of course, affirmations in *Zarathustra*; but they seldom, if ever, amount to more than proclaiming the need for a higher, stronger, more joyful human type. One seeks in vain for even the faintest suggestion of whether and how such an affirmation can ever replace the traditional philosophies of hope.

Let us take another example of how Nietzsche's "poetic method" enables him to shrug off important questions. In the section called "On the Higher Man" (Part Four, 13), Nietzsche's protagonist states that the first time he descended from the mountains and came to men, he committed the folly of standing in the marketplace. By the next day a new truth came to him: of what concern to him are the marketplace and the mob and the mob noise? The prophet urges higher men to learn this truth from him, that in the marketplace nobody believes in higher men. If they want to speak there, they had better prepare themselves for the mob's reply: there are no higher men, we are all equal, man is man, before God we are all equal. The prophet urges the higher men to stay away from the mob and the marketplace, reminding them that it is only now, when God lies in his tomb, that the opportunity exists for the higher men to rise. God is dead; now we want the superman to live.

Nietzsche, it appears, deliberately chose the "marketplace," the *agora*, as the foil for his "higher men" because he knew so well the role of the *agora* in classical Greece, and in Athens in particular. It was there that the citizens met to discuss politics. Instead of addressing the question, as did Socrates and Pericles, of what role the "people" should play in politics, Nietzsche simply settles the question by calling the denizens of the marketplace a mob. Yet, if we reflect on the history of Athenian or any other democracy, there is no evidence to support Nietzsche's implicit notion that the doctrine of equality necessarily thwarts the emergence of higher, creative types.

But it was not only in Athens but in ancient Israel as well that the *agora* constituted the context in which historically important encounters had occurred. For that was the context in which the Hebrew prophets delivered their message, admonishing the oppressors of the people. Knowing the Hebrew Bible intimately, as he did, would Nietzsche have denied that Amos, Hosea, Isaiah, et al., were "higher types"? Or does their concern for social justice automatically make them lower decadent types? And should these higher types have stayed away from the marketplace and the people, and have spoken only to themselves? We shall return in a later chapter to compare Zarathustra with the Hebrew prophets – this Zarathustra who has the superman at heart as his first and only concern, not man, not the neighbor, not the poor and most ailing, and not even the best of the all-too-human.

After many years of teaching (preaching?) his message, Zarathustra began to feel that his "virile nourishment" was taking effect. One day, however, when he was resting peacefully with his animals, and all the aspiring higher men were enjoying themselves as guests in his cave, he was startled by the sudden silence: their noise and joyful laughter had ceased. When he entered the cave to investigate, he could hardly believe his eyes: all of the aspiring higher men had become pious again, glorifying, adoring and praying to an ass who responded to each of their shouts of praise with "Hee-Haw!" They had created a veritable Ass Festival. When the prophet demanded an explanation, they replied, in effect, that it is better to adore God in this form than not at all. Naturally, the prophet reprimanded them for their childish foolishness and, pointing upward, acknowledged that except they become as little children, they shall not enter *that* Kingdom of Heaven. But then he reminded them once again that what he teaches has nothing to do with *that* kingdom. Zarathustra now realizes that these are not his proper companions; it is not for the likes of these that he waits in the mountains.

When the sun rose the next day he found himself surrounded by the animals he loved. Suddenly, a great throng of birds fluttered overhead, and then something strange happened: as he stretched out his hand, he unwittingly reached into a thick warm mane, and at the same time he heard in front of him a soft, long, lion's roar, which the prophet took as a sign. And, indeed, a mighty yellow animal lay at his feet, pressing its head against his knees out of love. Inspired and invigorated by the coming of the lion, the prophet exclaimed that this was his morning, his day. Renewing his resolution to continue his work, Zarathustra left his cave, radiant and strong as a morning sun that rises from behind dark mountains.

Ending on this note, Nietzsche has replaced the ass with the lion, which, to be sure, strikes a blow against theological asininities. But however much we may wish to appreciate Nietzsche's experiment in poetic philosophy, the basic question remains: can Nietzsche's lion-hearted affirmations of the superman ever fill the moral void created by his negations? An adequate reply to this question requires an examination of the later works in which his intellectual powers do in fact rise to a higher level.

NOTES

1 See A. V. Williams Jackson, *Zoroaster: The Prophet of Ancient Iran* (AMS Press Inc., New York, 1965); and Mary Boyce, *Zoroastrians* (Routledge & Kegan Paul, London, 1979).
2 *Varieties of Religious Experience* (Collier Books, New York, 1961), pp. 123, 139.
3 Walter Kaufmann, ed. and tr., *The Portable Nietzsche* (Penguin Books, New York, 1959), p. 347.

3

Beyond Good and Evil

Nietzsche was not entirely pleased with the way he had presented his thought experiments in *Zarathustra*. The "poetic" style had failed, in his judgment, to convey his ideas as clearly as he would have liked. *Beyond Good and Evil* was therefore designed to elaborate and clarify *Zarathustra*, just as *On the Genealogy of Morals*, published a year later (1887), was to perform the same service for *Beyond Good and Evil*.

Beyond Good and Evil maintains the strong Dionysian motif of *Zarathustra* with the proclamation of the will to power as the will to life, and the emphasis on the need to free the passions and instincts from the crippling influences of the old virtues. The central theme is the same as before; the old virtues have to be overcome, and this is the role and privilege of the very few who are very strong. One should shed the bad taste of wanting to agree with the multitude about the "good," and especially about the "common good," for whatever is common always has little value.

Nietzsche has no use for the modern philosophers among his contemporaries, the prolific scribbling slaves of the democratic taste who find the cause of all human misery in the social system. They foster in the herd a yearning for a green-pasture happiness, a new society offering security and comfort – a life free of risks and dangers. Catering to the masses, these democratic theorists speak incessantly of the "equality of rights"; they sympathize with all who suffer, and they dangle before the people the promise of abolishing all misery. All such doctrines are simply a translation of Christian eschatology into secular terms, with the consequent weakening of the human will to life. For the Christian faith was, from the start, a denial of the freedom, pride, and self-confidence of the human spirit. Like all religious neuroses, it has made dangerous and debilitating, anti-life demands – fasting, sexual abstinence, and other forms of self-denial. Christianity broke the strong and healthy by casting suspicion on the joy in beauty, and by devaluing the haughty, manly, conquering, and domineering qualities of the human being – all the qualities and instincts of the highest type of man. The Church has inverted these old noble values and has put in their place a suspicion and a hatred of the healthy human passions and instincts.

The will to power and life in the healthy and strong "higher type" presupposes that one's drives and passions are given free play, so that a man's sexuality reaches up into the heights of his spirit. One never feels as free and as euphoric as when one is at play. The real test of a man's maturity therefore consists in finding again the seriousness one had as a child at play. It is in music, above all other arts, that the passions and emotions enjoy themselves. The big mistake of the past, which continues to plague the present, is to look at the joy of the passions from a sickly moral standpoint. It is time to gain the courage to proclaim that the so-called "evil" passions are what is best in us.

Together with this salient Dionysian motif one finds a corresponding deprecation of reason. If philosophers were to look around themselves carefully, they would see that there are many moralities. When they claim to provide a "rational foundation for morality," they fail to recognize that what they have actually given us is merely a sophisticated variation of the common faith in the prevalent morality. On the one hand, then, Nietzsche has rejected metaphysics in all its forms as a criterion for evaluating human acts, relations, and products; and on the other hand, he also rejects reason. As Nietzsche sees it, philosophy still confronts the age-old problem of instinct *vs* reason. In the evaluation of things, he asks, which deserves more authority, instinct or rationality? And, of course, he favors Dionysus over Socrates. The latter, as a superior dialectician, had at first sided with rationality, striving always to answer the question "why" with good reasons. Indeed, Socrates derided the noble Athenians – who like all noble men were men of instinct – for their failure of his tests: they never managed to give good reasons for their actions. In the end, however, Socrates had to admit to himself, upon self-examination, that he too failed the test, demonstrating the same incapacities as his interlocutors. Socrates, according to Nietzsche, wanted it both ways; one must respond to the demands of the instincts, but enlist the aid of rationality in guiding them. That was the real error of that great ironic thinker; he satisfied his conscience with a piece of self-deception. He did, however, recognize the irrational element in moral judgments (191).

If Nietzsche rejects both faith and reason as means of grounding moral judgments, then his "will to power" is truly "beyond good and evil." "There are no moral phenomena at all," he writes, "but only a moral interpretation of phenomena . . ." (108). All right, let us go along with this for the moment and agree that the will to power and giving vent to the passions are not, in and of themselves, moral phenomena. Are they nevertheless subject, for Nietzsche, to a moral interpretation of some kind? If they are, how would Nietzsche morally evaluate a phenomenon without reason? And if they are not subject to moral interpretation, does that mean that for Nietzsche *anything* one does to realize one's will is permissible? Further, if Nietzsche criticizes Socrates' attempt to guide the passions and instincts by means of reason, does Nietzsche intend to imply thereby that the instincts and passions should be left alone to act blindly? To put the question in Freudian terms, does Nietzsche advocate that the natural drives – the *id* – be totally unconstrained? It is doubtful in the extreme that he advocates such a thing, since a totally unconstrained *id* is either an impossibility in the human condition, or a temporary, destructive aberration that no society would tolerate for long. So on the one hand Nietzsche often speaks as if he wants to free the instincts from all social controls, and on the other he speaks of "self-mastery" as his highest ideal. What element is it, in the psyche of the human being generally and of the "higher type" in particular, that accounts for "self-mastery" if it is not reason? Though Nietzsche does not enlighten us as to the nature of that element, we might justifiably guess that it is an Apollinian faculty of the psyche – that is, an aesthetic faculty that enables us to distinguish between balance, harmony, moderation, etc., and their opposites. And if that is the sole regulator of the will to power, then whether it is properly regulated or not is, for Nietzsche, strictly a matter of *taste*. Finally, as we have already observed, Nietzsche ignores the Hobbesian problem. If power refers to one's present means to obtain some future apparent good, what would Nietzsche propose with respect to a situation in which two or more individuals strive for the same good which, nevertheless, they cannot both have? Nowhere in his writings does Nietzsche face this problem; and if he provides neither a rational nor a moral means of resolving a conflict of wills, then all we are left with is "might is right." Add to all that his selection of Cesare Borgia and other such ruthless historical persons as "healthy types," in whom there was nothing pathological, and we can easily see how Nietzsche allows himself to be thus interpreted.

Nietzsche's dichotomy of the "higher type" and the "herd" also opens the door to such an interpretation. It has always been true, he reminds us, that the few have commanded while the many have obeyed. Obedience has for so long been cultivated in the masses that it has become an innate need. Indeed, the herd-instinct of obedience is inherited most easily, and at the expense of the art of commanding. If this were to continue to its ultimate phase, the strong spirits who command would disappear altogether, or they would catch the herd disease, suffer from a bad conscience, and proclaim that when they command they actually obey. They deceive either themselves or others or both, by posing as the executors of higher commands – of tradition, of the constitution, of right, and even of God. They borrow maxims, such as "servants of the people," or "instruments of the common weal," from the herd's way of thinking. Even when the need for real leaders is recognized as indispensable, strenuous efforts are made to replace commanders with herd-men. That is how parliaments originated. But the truth is that the herd-animal Europeans experience great comfort and even salvation when a "higher type" commander appears. "The history of the response to Napoleon," writes Nietzsche, "is almost the history of the higher happiness conveyed by this entire century through its most valuable human beings and moments" (199). Napoleon was a civilized man by the standards of his era, and certainly by the standards of twentieth-century "strong men." But one can easily see how Nietzsche's admiration for the strong "higher type", together with his antipathy for democracy and his theory of the will to power as beyond good and evil, can be viewed as a theory that "might is right." Nietzsche's examples from antiquity only strengthen the grounds for such an interpretation, since the men he singles out for admiration are Alcibiades and Julius Caesar.

Let us pause for a moment and take a close look at this Alcibiades whom Nietzsche so much admires. During the Peloponnesian War the democratic government of Athens decided that the conquest of Syracuse was necessary as a key to Athens' power in the West. The privileged classes and the pro-oligarchical elements opposed the whole enterprise. They appointed Nicias, a respected general and a decent man, as their spokesman. He sincerely believed that the proposed Syracusan expedition was misguided, foolhardy, and doomed to failure. In an attempt to discourage the enthusiastic supporters of the adventure, Nicias argued that the cost they had estimated was much too low, and that the Athenian Assembly should double its allocation. We learn from Thucydides (VI, 20–25)[1] that the Assembly, ironically, was so delighted with Nicias' caution that they named him, together with two others, to the staff of generals in charge of the expedition. The two other appointees were Lamachus, a serious and business-like military man, and Alcibiades. Though brilliant, Alcibiades' political career up to this point had been vacillating and unsure. But he had been close to the circle of Pericles, and the Assembly therefore hoped that he would be reliable. While preparations for the expedition were going on, it was found that in one night nearly all the stone statues of Hermes in the city of Athens had had their faces disfigured. Hermes was the god of the merchant class, and in some ways a sacred symbol of the democratic party. On the discovery of the mutilation one of those accused was Alcibiades, who, however, denied the charges. The democrats now had second thoughts about going ahead with the Syracuse expedition, since two of the generals, Nicias and Lamachus, seemed to have oligarchic leanings and the third, Alcibiades, was under suspicion. But after deliberation it was decided that the expedition should sail. Lamachus proposed an immediate and direct attack on Syracuse before the Syracusan forces would have a chance to mobilize their defenses. Nicias disagreed, calling instead for a show of force around the coasts of Sicily, where Syracuse was located. Alcibiades wanted a compromise of sorts; but at this point the Assembly voted that he should be arrested and recalled to stand trial on the charge of having participated in the mutilation of the statues of Hermes. He was in fact arrested, but on his return journey succeeded in eluding his captors and, to the astonishment of the democrats, turned up in, of all places, Sparta.

Evidently, he managed to ingratiate himself with the Spartan rulers by describing democracy as foolishness and by giving them practical proposals for the conduct of their campaign, based on his firsthand knowledge of the Athenian forces, their strengths and weaknesses. It was Alcibiades who suggested to the Spartans the fortification of Decalaea on the northern boundary of Attica so as to intercept Athenian trade with the North. That being not enough, he also proposed that a capable Spartan general be dispatched to Syracuse to assist in the organization of that city's defenses against the Athenian invasion. The person chosen, Gylippus, did the job so well that the besieging Athenian fleet was itself bottled up and besieged within the great harbor of Syracuse. With all attempts to break out failing, the Athenian commanders resolved to make a last desperate attempt to escape overland. Owing to the eclipse of the moon, however, and the religious significance this apparently had for Nicias, the departure was postponed until the full moon. An opportunity was lost (Thucydides VII, 50). In a dreadful scene on the Assinarus River, described in vivid detail by Thucydides (VII, 72–87), the beaten and despairing fragment of the once invincible Athenian armada was cut to pieces by the pursuing Syracusan forces. Poor Nicias himself was killed, and thousands of Athenians were taken into captivity and subjected to forced labor in the stone quarries of Syracuse.

Meanwhile, the irrepressible young Alcibiades was building himself a political career in Sparta; but unfortunately for his new ambition he had seduced the wife of King Agis, and therefore had to leave Sparta in something of a hurry. We next hear of him at the court of Persia, giving advice to the Persian King on how best to intervene in Aegean politics in order to advance the interests of the Persian Empire. However, after a complex chain of events Alcibiades is eventually restored to power in Athens where he performs his "first great act of service to his country" (Thucydides, VIII, 86f). It is indicative of Nietzsche's general outlook that he should admire this admittedly brilliant, strong-willed, politically agile, but apparently unscrupulous young man.

Nietzsche was aware that his provocative use of "herd" and "herd instinct" might be charged against him by the men of "modern ideas", but he maintained that these were appropriate epithets that expressed exactly his novel insight: what European morality calls "good" is nothing other than the instinct of the herd-animal in man, which has prevailed over other instincts. But prior to the emergence of this morality on earth, an older, higher morality was predominant. Here Nietzsche has in mind the noble "master morality" that had already in antiquity been inverted by and fused with a "slave morality". To understand how this inversion took place, we have to wait for *On the Genealogy of Morals* where Nietzsche for the first time presents his thesis systematically and coherently. Since Nietzsche believed a "higher" morality had once held sway, it ought to be possible, he assumed, to re-establish a noble morality *after* the herd-morality has been extinguished once and for all. The task would be far from easy, however, in the face of the huge stumbling blocks standing before the new master-morality – Christianity and its direct heir, the democratic movement.

For Nietzsche, democracy is not only a decadent political movement, but a pronounced diminution of man, which lowers his value and turns him into a mediocrity. New philosophers must therefore arise who are sufficiently strong and original to begin the process of re-inverting the herd values and creating higher values opposed to those of the herd. "Higher?" What ground does Nietzsche have for calling either the pre-herd or the post-herd morality higher? Nowhere does he provide any such ground, nor can he, since he claims to have rejected both metaphysical and rational means of doing so. Striking a Platonic note that sounds like a yearning for philosopher-kings, Nietzsche calls for new philosopher-*commanders* (203), in comparison with whom all previous philosophers will look pale and dwarfed. In the same context, however, there is a conspicuous sentence in which Nietzsche acknowledges, in effect, that his expectation of the coming of those new leaders is neither more nor less than an act of *faith*: "the necessity of such leaders, the frightening danger that they might fail to appear or that they might miscarry or degenerate – these are *our* real worries and gloom . . ." (203). So it turns out that Nietzsche the materialist, who has rejected all metaphysics and in particular the faith propagated by the Church, has a faith of his own.

Even modern science, with its declaration of independence from philosophy and its claim to objectivity, is viewed by Nietzsche as a consequence of the democratic movement. In demanding freedom from all masters, scientists, perhaps unwittingly, express the instinct of the rabble. After emancipating itself from theology, whose "handmaid" it was too long, science now wants to lay down laws for philosophy. On the one hand, Nietzsche welcomed the movement towards objectivity in both science and scholarship; on the other hand, he also saw drawbacks. For the objective spirit, especially among the "positivist" philosophers, does after all result in an "unselfing" and depersonalizing which, for Nietzsche, tends to undermine the philosophic enterprise as he conceives it. Here Nietzsche was among the first to make an all-important observation: science is only an instrument. It is neither an end in itself, nor can it create ends or values. Science requires the guidance of philosophy, but the "positivist," who sees himself as authentic only when he is objective, no longer knows how to affirm or negate, how to take sides. In honoring objectivity, then, philosophers have overlooked the fact that science can only provide us with means, not ends. The objective man is a precious instrument for measuring who deserves care and honor, but he is no goal.

Nietzsche sees a connection between objectivity and skepticism, the latter being an attitude that develops when peoples and classes which have long been separated are brought together and crossed. The new generation that emerges from such mixing thus inherits diverse and conflicting standards and values, which create unrest and doubt. In hybrid individuals of this kind, the best forces within them tend to inhibit and counteract one another, preventing the real virtues from growing and becoming strong. Such individuals, pulled, as they are, first to one side and then to another, lack a center of gravity and perpendicular poise. Most severely weakened in the skeptic is the *will*: he has lost both the capacity for independent thinking and the pleasure in willing. Skepticism is a fence-sitting paralysis of the will, which expressed itself in Nietzsche's era in such fads and fashions as "objectivity," "scientism," and "pure knowledge, free of values and will."

For Nietzsche, skepticism, as the weakening of the will, was most pronounced in the so-called civilized nations of Europe; and it tended to decline to the extent that the "barbarian" impulse in Western culture asserted itself. Strength of will was therefore strongest in that enormous empire in which Europe flows back into Asia – Russia. There, the accumulation of will, which has proceeded for centuries, has created so strong a will – whether to affirm or deny it itself does not know – that it waits menacingly to be discharged. To rid Europe of its greatest menace, Russia, internal upheavals would be necessary, leading either to the breakup of the Russian empire into small units, or to the destruction of

absolutism through an introduction of what Nietzsche calls the "parlia-
mentary nonsense." Now, it is not entirely clear whether Nietzsche
intends the very next sentences to be taken literally, but given his zealous
concern for the strengthening of the European man's will, one cannot
simply discount such utterances:

> I do not say this because I want it to happen: the opposite would be closer
> to my heart – I mean such an increase in the menace of Russia that Europe
> would also have to decide to become menacing, that is, to *acquire one will*
> by means of a new caste that would rule Europe, a long terrible will of its
> own that would be able to set its goals for millennia hence – so the
> protracted comedy of its numerous petty states as well as its dynastic and
> democratic factions would come to an end. The time for petty politics
> is gone: already the very next century will bring the struggle for the
> dominion of the earth – the *compulsion* to politics on a grand scale. (*den
> Zwang zur grossen Politik* (208)

It is utterances such as this that made it easy for Nazi theorists to
appropriate Nietzsche and to adapt his philosophy to their ends.

Nietzsche insists that one must stop confounding critics, philosophical laborers and scientists with real philosophers. "*Genuine philosophers*,"
writes Nietzsche,

> . . . *are commanders and legislators*: they say, "thus it *shall* be!" They first
> determine the Whither and For What of man, and then make use of the
> preliminary work of all philosophical laborers, all who have overcome the
> past. Reaching for the future with a creative hand, everything that is and
> has been becomes for them an instrument, a hammer. Their "knowing" is
> creating, their creating is a giving of laws, their will to truth is – a will to
> power. (211)

In such passages too, one hears a loud authoritarian chord. If Nietzsche's
"new" philosophers are to be "commanders and legislators," then we
are justified in suggesting that his vision, in this respect, is quite
similar to that of Plato in the *Republic*, whose philosophers were to
hold a monopoly of political power. We must remember that Plato's
philosopher-guardians were to be a ruling class in the full sense of that

term, in that Plato gave them a monopolistic control over the means of violence. In contrast, the largest class in Plato's scheme – the "producers" or "farmers" – were assigned no political role whatsoever. And since Nietzsche shares Plato's strong antipathy for democracy, one may reasonably infer that when Nietzsche speaks of the genuine philosophers as commanders whose will to truth is a will to *power*, he too, envisages the philosophers as kings who, like Plato's Guardians, will also possess the physical means of realizing their will against the resistance of the rest of the populace. That Nietzsche's concept of power implies domination is also borne out by the contrast he draws between the old philosophers and the new, genuine ones whom he emvisions: in the past the task of philosophers amounted to little more than being the bad conscience of their time, but the genuine philosophers of the future will embody such greatness that they will be "beyond good and evil" (212). Walter Kaufmann has commented in this regard that "the element of snobbery and the infatuation with 'dominating' and 'looking down' are perhaps more obvious than Nietzsche's perpetual sublimation and spiritualization of these and other similar qualities."[2] As we have seen, however, and as we shall continue to see, it is rather doubtful whether Nietzsche's theory can be interpreted as strictly spiritual and without any political implications.

It is true that Nietzsche views the prevailing morality as a form of spiritual revenge against those who are less limited. Moral judgments are malice spiritualized. This is the central idea that he systematically develops for the first time in his *Genealogy*, to be considered in the next chapter. But it is also true that Nietzsche sees an order of rank in individuals; the higher and rarer types should command while the lower ranks obey. That is why he rejects political equality; for it is *immoral*, he says, to suppose that what is right for the one is also right for the other. Europe, he says repeatedly, has been plunged into a semi-barbarism by the democratic mingling of classes and peoples. In this respect, too, Nietzsche's views resemble those of Plato in the *Republic*. But Plato's Socrates gives reasons for his rejection of democracy, namely, that politics, like every art, requires specialized knowledge which, he argued, the common Athenian citizens did not possess. Nietzsche, in contrast, never examines the merits of democracy as compared with other systems. Indeed, he cannot engage in such an examination, since the nature of his philosophical outlook is so overwhelmingly Dionysian that it has crowded out the Socratic, rational element.

And this is true throughout. Take, for example, the English Utilitarians, whom Nietzsche describes as "ponderous herd animals" with bad consciences. These thinkers advocate the cause of egoism as the cause of the general welfare, arguing that if each individual pursues his interests relentlessly, it will lead to the greatest good for the greatest number. Now it is doubtful whether the market system has ever worked in the automatic, self-regulating way which the classical economists and other Utilitarians imagined; and it is even more doubtful whether it worked for the general good. As early as the second decade of the nineteenth century, the Swiss economist, Sismondi, in his *Nouveaux Principes d'Economie Politique* (1819), demonstrated that the poor suffer more from economic crises, and that the Utilitarians were therefore simply wrong. But Nietzsche has no interest in the question of whether the utilitarian theory is valid or not. What annoys him about these English theorists is their concern for the "general welfare," which, for Nietzsche, is certainly no goal but rather an "emetic." He berates these thinkers for their failure to recognize that the "demand of one morality for all is an injury (*Beeinträchtigung*) to the higher men; that there is an order of rank between man and man, consequently also between morality and morality" (228). In this fairly typical way Nietzsche leaves the matter with an unsupported assertion, and with no explanation of why that should necessarily be the case.

Another example is Nietzsche's attitude towards women. Nietzsche sees a fundamental antagonism between man and woman, an eternally hostile tension. One must think about women as the Orientals do, as property that can be locked up and as creatures predestined for service. Thinkers who advocate equal rights, equal education and equal obligations for women are, therefore, too shallow to serve as the higher philosophers of a future era. In no age have women been accorded as much respect by men as in the present democratic age, which proves that the notion of emancipating woman is yet another product of the dominant herd mentality. Plato, no democrat, admitted women to the ranks of the philosopher-guardians if they met the qualifications; and he explained what a serious loss it would be to society if those ranks were closed to women. Nietzsche, in contrast, merely asserts dogmatically that the first and last profession of women is to give birth to strong children (231–9).

Aristocracy, for Nietzsche, has always embodied what is noble in men, and will do so again in the future. The aristocratic societies have brought out the best in man because they believed in an order of rank and in differences of value between one man and another, and because they also believed in the necessity of slavery in one form or another. However, one must not have illusions about aristocratic societies and the nature of the men who constituted them: they were barbarians in the terrible sense of the term, predators who possessed an enormous strength of will and an insatiable lust for power, and who threw themselves upon weaker, more civilized and peaceful peoples. In the early periods of history, the noble caste was always the barbarian caste whose predominance rested not only on their physical strength and prowess, but on their definition of themselves as the "good." The chief characteristic of a healthy aristocracy is that it experiences itself as fully justified to rule and dominate. It therefore accepts with a good conscience

> the sacrifice of an untold number of human beings who, *for its sake*, must be lowered and turned into incomplete human beings, into slaves and instruments. Their [the aristocrats'] basic faith has to be that society exists not for its own sake but only as the foundation and scaffolding on which a select type of being is able to raise itself to its higher task and, in general, to a higher state of *being* . . . (258)

Among aristocrats – that is, higher men who are similar in strength and in value standards – it may be good manners to refrain mutually, from injury, violence and exploitation; but as soon as such manners are extended to the society as a whole, thus becoming its fundamental moral principle, they result in a denial of life and in deterioration and decay. Exploitation, injury and violence are not manifestations of imperfect, corrupt or primitive societies; they are the essence of anything that lives – consequences of the will to power, which is, after all, the will to life. And to understand the will to life realistically, one must resist sentimentality and weakness: life is, in its essence, the appropriation, injury and overpowering of what is "alien" and weaker.

Here as elsewhere Nietzsche entertains a false interpretation of

Darwin's theory; and although we shall discuss Nietzsche's error at length in Chapter 9, a few words are in order here. When Nietzsche speaks of the cruelty of life and the will to overpower, pointing to nature, as it were, to support his contention, he draws his examples from the *inter*-species behavior of animal life in which one species is predator and the other prey. This is inappropriate and misleading, as we shall see, since what Darwin has to say about the *intra*-species behavior of animals tends to contradict Nietzsche's view. The conflicts between the stronger individual members of a species, Darwin demonstrated, produce the best leaders; and their leadership, in turn, enhances the co-operative character of the social organization of that species. The more co-operative the species, the greater is its adaptability. Hence there is no dichotomous or antithetical relationship between the "stronger type" and the "herd," as Nietzsche imagined. On the contrary, the role and function of the stronger types is to enhance the adaptive capacity of the species as a whole. And while it is true, of course, that exploitation, injury and violence are facts of human history, it does not follow that, as Nietzsche asserts, they are beyond good and evil.

Beyond Good and Evil is the work in which Nietzsche had first introduced his now famous terms, "master morality" and "slave morality," which he explicated more fully in the *Genealogy*. These key terms are the pure or ideal conceptual constructs with which Nietzsche seeks to explain how the ideas of "good" and "evil" had first emerged – the origins of morality. In what is undoubtedly Nietzsche's major contribution to our understanding of such origins, he proposes that the moral distinction of values first appeared both in a dominant group – whose awareness of its difference from the dominated group was accompanied by pride – *and* among the dominated, slaves and dependents of all kinds. When the ruling group determines what is "good," that term refers to its own powerful and exalted status, to its superiority over the servile lower orders who are despised. In this "master morality" the opposition of "good" and "bad" means roughly the same as "noble"and "contemptible." The dominant, noble man, who is, above all, a courageous man of war, feels contempt for the dominated and servile because they are, in his view, fearful and cowardly, allowing themselves to be maltreated. The noble type of man experiences himself as the creator of values who needs only his own approval and that of his peers. What is harmful to him is harmful itself; all power-enhancing aspects of his life are honorable. Noble morality is a form of unabashed self-glorification. Life for the noble man is the feeling of the fullness of power, the joy in adventure and high tension, the awareness of wealth that one can bestow on one's noble heirs. The noble human being also helps those who are lowly and unfortunate, but this is prompted not by pity but primarily by his super-abundant power: *noblesse oblige*! "Master morality" and "slave morality" are, as we have said, pure, antithetical, intellectual constructs. Nietzsche emphasizes, however, that in reality these moralities are never found in their pure form, but are rather mixed. In all cultures one may discern attempts at mediation between the two moralities, and often an interpenetration of both. The two moralities may even co-exist in the same human being, within a single soul (260).

In the *Genealogy*, where Nietzsche approaches the origin of the two moralities historically, he will characterize "slave morality" as an *inversion* of noble values, a product of the slaves' resentment of their oppressors, a form of spiritual revenge. In *Beyond Good and Evil*, however, he has not yet reached that level of clarity, so what we find there are his first thought-experiments on the nature of "slave morality." When the suffering, oppressed, and unfree moralize, they express a pessimistic and condemnatory view of the human condition. They do not look favorably upon the self-glorifying virtues of the powerful, and they are skeptical and suspicious of the so-called "good" that is honored there. This gives us insight into the origins of the opposition of "good" and "evil." The slaves' experience of their oppressors as powerful, danger-ous, cruel and fear-inspiring, gives birth to the idea that the oppressors are the embodiment of "evil." However, the masters' experience of their exalted and dominant status engenders in them the unshakable convic-tion that it is precisely those who inspire fear, and wish to inspire it, who are the "good." The fearful and weak are "bad," that is, contemptible. Finally, and not surprisingly, the longing for freedom and the feeling of happiness it brings belong necessarily to the slave morality.

The noble virtues of strength, courage and hardness emerge of necessity from the life-circumstances of the masters who have to stick together if they want to prevail. They must prevail or run the risk of being destroyed, either by hostile neighbors or by their own oppressed masses. Experience thus teaches the noble warrior caste to which qualities it owes its successes and triumphs, and it is those qualities which are cultivated and called virtues. The noble soul is above all egoistic, possessing the conviction that he and his peers are superior, and that those other beings, who are inferior by nature, must subordinate and sacrifice themselves. Nietzsche, of course, favors and admires the master morality whose noble qualities he hopes to see in the "higher men" of the future. In *Beyond Good and Evil* there is no proclamation of the "superman." Indeed, as compared with *Zarathustra*, that theme is conspicuously absent. There is, however, one aphorism in which he speaks of the "problem of those who are waiting," and he says:

> Strokes of luck and much that is incalculable are necessary if a higher man, in whom the solution of a problem lies dormant (*schläft*), is to get around to action in time . . . (274)

So the coming of the "higher man" is a matter of luck, and Nietzsche's expectation in that regard remains an act of faith. He continues to believe that where the solution of problems is concerned, Socratic, dialogical reasoning is useless.

Nietzsche had an obvious distaste for what he calls "slave morality" and an equally obvious admiration for "master morality." Does his admiration imply endorsement? That question prompts us to ask another: how would Nietzsche decide whether to endorse one morality over another? If he has rejected both transcendental and rational criteria, as seems to be the case, what other criterion remains by which to choose between moral values? Or would Nietzsche say that in asking such questions we have missed the whole point of his philosophy and, in particular, of this book – namely, that it is an illusion to suppose that any such real criteria exist at all? If that is Nietzsche's position, it follows that his preference for, say, the *Iliad* (master morality) over the New Testament (slave morality) is no moral preference at all, but strictly a matter of taste – *his* taste. *De gustibus non est disputandum*!

Nietzsche concludes *Beyond Good and Evil* by paying homage once again to the god Dionysus, whom he has so much admired from the time of *The Birth of Tragedy*. There, in his first book, Dionysus represents frenzied, intoxicated passion that requires the balance and order of Apollo. But in Nietzsche's later works, including the one under consideration, there are only the faintest traces of an Apollinian element, while the Socratic element has altogether disappeared. One wonders how, in the absence of those elements, Nietzsche supposed that the Dionysian passions would be harnessed and creatively employed.

NOTES

1 *History of the Peloponnesian War*, tr. Rex Warner (Penguin Books, London, 1954).
2 Friedrich Nietzsche, *Beyond Good and Evil*, ed. and tr. Walter Kaufmann (Vintage Books, New York, 1966) p. 140, n. 37.

4

On the Genealogy of Morals: Ressentiment and the Inversion of Values

Nietzsche described *Beyond Good and Evil* (1886) as an elaboration and clarification of the views he expressed in *Zarathustra*, and *On the Genealogy of Morals* (1887) as performing the same task for *Beyond Good and Evil*. It is in the *Genealogy* that we find the most coherent, systematic and lucid account of Nietzsche's theory of ethics. The *Genealogy* is a brilliant and original work in which Nietzsche most convincingly employs psychology, philosophy and classical philology in order to trace Western values to their roots. His theory of the origins of good and evil has proved to be quite fruitful in sociological studies of religion; but, as we shall see, that does not mean that the conclusions Nietzsche himself drew from his theory are philosophically valid.

The basic question addressed in the *Genealogy* is where and how the Western concepts of good and evil had originated. The method Nietzsche employs is historical-sociological in that he asks this question: what were the social conditions in which human beings first formed value judgments of good and evil? This question is treated separately from the philosophical question of what value those judgments of good and evil actually possess. Nietzsche's criterion for assessing the value of "values" is whether they have furthered or hindered the will to life. Do the prevalent values promote the impoverishment and degeneration of life, or, on the contrary, the strong and courageous fullness of life?

So Nietzsche's first task is to provide an analysis of the social circum-stances in which Western values emerged and developed; and of course in providing such an analysis Nietzsche intends to call into question the value of those values themselves. His aim, in a word, is to supply us with a history of morality, and in so doing to challenge the validity of that morality. When we say a "history" of morality, this should not be taken to mean that Nietzsche gives us a detailed historical account of either the origin or the inversion of values. He himself only hints at the circumstances in question, so we shall have to complement and expand his account with our own details from the history of ancient Israel and ancient Greece, the two cases with which he deals most directly.

Nietzsche begins with the question of what the concept and judgment of "good" had meant originally. If someone had no knowledge of the history of the concept "good," he might suppose that actions were originally approved of and called good because they were unegoistic and because those to whom they were done found them useful. But this supposition is unhistorical, and fundamentally wrong. Here Nietzsche introduces a truly original insight: the judgment "good" did not originate with those to whom actions were done, but, on the contrary, with the good themselves, that is to say with the noble, powerful, high-stationed and high-minded who looked upon themselves and their actions as good in contradistinction to the low, common, and plebeian. It was the experience and feeling of the nobility, as the dominant and higher ruling order in relation to a lower order, that first gave rise to the antithesis "good" and "bad."

In the history of Greece, for example, the gradual breakdown of the primitive monarchies turned to the advantage of the powerful noble-warrior chiefs, who became the masters of the city-states and remained so for centuries. Tracing their origin to some deity and taking immense pride in their noble blood, they fastidiously preserved their genealogical tree and the traditional history of their house (*patria*). As the leaders of powerful clans they controlled land and revenues of sizeable domains and enjoyed the riches won at the point of the sword over many generations. Throughout Greece a class of noblemen emerged, designated by such terms as "the good" (*agathoi*), "the best men" (*aristoi, beltistoi*), "the great and good" (*kaloi kagathoi*), "men of blood" (*eugeneis, gennaioi*), "men of quality or truth" (*esthloi, chrestoi*), "men of honor" (*gnorimoi, epieikeis*). They were also called "well-born men," "lords of the earth," and "knights."

It follows from the noble origin of the word "good" that originally it had nothing to do with "unegoistic" actions. When, then, did "good" and "unegoistic" become linked? Only with the decline of the nobility. For Nietzsche, a historical approach therefore makes it plain that originally "noble" or "aristocratic", in the social sense, was the basic source from which "good," in the sense of possessing a soul of a higher order, had developed. And this process ran parallel to another in which "common," "plebeian," and "low" were transformed into the concept "bad." From Nietzsche's philological studies he also gained the insight that although the Greek nobles most often designated themselves by their superiority in power, they also called themselves "the truthful." That is how they were described by the Megarian poet Theognis who wrote in the sixth century BC. The root of the Greek word for this, *esthlos*, denotes one who possesses reality, one who is actual or true. This is then given a subjective turn so that it is exclusively the nobleman who tells the truth in contrast to the common man who lies – which is how Theognis, again, describes him. In the Greek word *kakos* (bad), as in *deilos* (cowardly), cowardice is also emphasized, which indicates where one should look for the etymological origin of *agathos* (good, well-born, brave).

Similarly, when the highest stratum in society is the priestly caste, the concept denoting political superiority carries with it the connotation of the superiority of soul. Then "pure" and "impure" enter the picture as designations of social station. But "pure" from the outset merely referred to those who washed themselves, who forbade and avoided certain foods that caused skin and other ailments, who did not sleep with women of the lower strata, and so on. Although the priests as a caste may have roots in the noble strata, the priestly mode of evaluation can develop into the antithesis of aristocratic values, and this is highly probable when the priests and noble warriors stand in zealous opposition to one another and refuse to compromise. The knightly-noble values presupposed great health and a powerful physicality required for war, plundering expeditions, hunting and anything else that entailed vigorous, adventurous and joyful activity. The priestly-noble mode of evaluation, owing to the specialization of the priests in "sacred" functions, soon turns in another direction. As a caste they lose their physical prowess and war skills, becoming relatively weak and even powerless. Priests, for Nietzsche, are the most bitter enemies of anyone they oppose, and it is precisely because of their impotence that their hatred and vengefulness reach monstrous proportions. Nietzsche calls the Jews a priestly people *par excellence*, and describes in the most dramatic terms the historic role of the Jews in inverting the noble values:

> All that has been done on earth against "the noble," "the powerful," "the masters," "the rulers," is not even worth talking about when compared with what the *Jews* have done against them. The Jews, that priestly people, who in opposing their enemies and conquerors gained satisfaction only through a radical revaluation of their enemies' values, that is to say, through an act of the *most spiritual revenge*. . . . It was the Jews who, with awe-inspiring consistency, dared to invert the aristocratic value-equation (good = noble = powerful = beautiful = happy = beloved of God) and to cling to this inversion with their teeth, the teeth of the most abysmal hatred (the hatred due to impotence), establishing the principle that "the wretched alone are the good; the poor, powerless, lowly alone are the good; the suffering, deprived, sick, ugly, alone are pious, alone are blessed by God – blessedness is for them alone; and you, the powerful and noble, are, on the contrary, the evil, the cruel, the lustful, the insatiable, the godless to all eternity; and you shall be for all eternity the unblessed, accursed, and damned!". . . . With regard to the tremendous and im-measurably fateful initiative which the Jews have taken, through this most far-reaching of all declarations of war, I recall the proposition I arrived at on an earlier occasion (*Beyond Good and Evil*, 195) – that *the slave revolt in morality* begins with the Jews, a revolt which has a two-thousand-year history behind it and which is no longer so obvious because it has been victorious. (*Genealogy*, I, 7)

In the history of the West it was ancient Israel's resentment of her Egyptian oppressors that first gave rise to a form of spiritual revenge, an inversion of the noble values and a triumph over them. The "slave morality" of Moses and the prophets had triumphed over the values of the high and mighty.

Nietzsche drew from this historical fact a relativistic conclusion: since the inversion of values occurred in specific historical circumstances (which we shall soon describe in some detail), the ethic taught in the Hebrew Bible (the Old Testament) is contingent on those circumstances. This apparently meant, for Nietzsche, that the Hebrew ethic can make no claim to a lasting, trans-historical validity. In drawing such a conclusion Nietzsche can be accused of having committed the "genetic fallacy," the false notion that the social origin of an idea has necessary implications for the validity of that idea. For example, the Hebrew conception of justice emerged in a particular context of ancient Israel's history. Does that necessarily mean that this conception of justice has no validity beyond that context? Nietzsche proceeds throughout as if the ethical and moral teachings of Judaism and Christianity are strictly contingent and relative. Those teachings, moreover, together with the Socratic-Platonic legacy, are viewed as the ultimate sources of the decadent and degenerate character of western culture. Democracy can be traced to that source, which means, for Nietzsche, that the "people," the "slaves," the "mob," the "herd," or whatever one chooses to call them, have won.

How, according to Nietzsche, does the slave revolt in morality give birth to countervalues? It does so through what Nietzsche calls *ressentiment*, a psychological process by which the weak, denied the possibility of a reaction against the strong in the form of deeds, compensate themselves with an imaginary or spiritual revenge. *Ressentiment* entails a negation and repudiation of the master's values, a saying of No!, which eventually becomes the creative act of inverting those values and substituting new ones for them. The new values arise out of opposition to a hostile, oppressive, external world. The psychological experience of *ressentiment*, if it appears at all in the noble individual, soon consumes itself in an overt reaction against the adversary and therefore does not *poison*; in the weak and impotent, in contrast, *ressentiment* converts the enemy or oppressor into the "evil one." In the morality of *ressentiment* the "good man" of the other morality, that is, the noble, powerful man or ruler, is transformed into the source of evil. Nietzsche does not of course deny that from the standpoint of the weak and oppressed, who know the noble and powerful only as enemies, there is good reason to regard them as evil. He recognizes that mutual respect, suspicion and jealously hold the noble warriors in check in their relations with one another, but not in relation to outsiders. Against strangers and enemies, the "good" and the "noble" become enraged beasts of prey who find their feats of murder, arson, rape and torture exhilarating, and who take pride in the new material they have provided for the poet's song and praise. Nietzsche compares the attitude of the men of *ressentiment* – and we encountered this metaphor earlier in *Zarathustra* – to that of lambs who dislike great birds of prey, and who call them evil and themselves good. The "lambs" are the majority of mortals, the weak and the oppressed of every kind, clad in the ostentatious garb of virtue, as if their weakness were willed and chosen and as if weakness were meritorious. In the *Genealogy* as in his earlier writings, Nietzsche makes no attempt to assess the validity of the "inverted" ethical teachings, but merely asserts that they are manifestations of weakness, impotence and decadence.

For Nietzsche, humanity has been engaged in a fearful struggle for thousands of years, a struggle between two opposing value systems: "good and bad" *vs* "good and evil." From the time of the Roman Empire the symbol of this struggle has been "Rome against Judea, Judea against Rome"; and no event has been more significant than this deadly confrontation. Rome looked upon the Jew as a dangerous antipode, and rightly so, says Nietzsche, ". . . provided one has a right to regard the future salvation of the human race as contingent upon the unconditional dominance of aristocratic values, Roman values" (*Genealogy*, 16). It is such passages that suggest an interpretation different from that of Walter Kaufmann who maintains that Nietzsche's conceptions of the "will to power" and the "higher specimens" refer primarily to the Goethes, Shakespeares and Beethovens of this world. If Nietzsche links the salvation of the human race to Roman values, and if those values are "beyond good and evil," then he does, in effect, espouse the doctrine, "might is right." The Romans were the mighty and noble, says Nietzsche, and ". . . nobody mightier and nobler has yet existed on earth or even been dreamed of . . ." (*Genealogy*, 16). Power and might do, then, appear to be the paramount values which Nietzsche admires and which the *ressentiment* of the Jews, that people of "unequaled popular moral genius," has negated and inverted. Which of these, Rome or Judea, has won for the present? The answer, for Nietzsche, is beyond doubt. Consider to whom one bows down in Rome itself, and not only in Rome but over half the earth. For Nietzsche, there was considerable truth in the proposition that from the time of Moses' repudiation of the values of the Egyptian "house of bondage," Judaism developed as a negation and inversion of the oppressors' ideals. And Jesus' repudiation of force and violence in Matthew 5: 38*ff* may also be understood in that light. Jesus' teachings may be viewed as a continuation and accentuation of the inversion process, in his case, a rejection of Roman (i.e. pagan) ideals of war, power and might. Thus three Jews, says Nietzsche, Jesus, Peter, and Paul proclaimed the countervalues that led to the victory of Judea over Rome.

For a better understanding of how Nietzsche perceived the Christian inversion of Roman values, we need to supplement his own remarks with a brief sociological sketch. How shall we understand the New Testament saying, "The last shall be the first"? For Nietzsche this can only be understood if we become aware of the significance of resentment in the formation of moral judgments. An analysis of the socio-historical context in which the sentence was first uttered and the idea first emerged suggests that it had a strong appeal for those who, like the early Christians, were oppressed and who, under the impulse of resentment, wished to liberate themselves from Roman domination. The circumstances of the oppressed engendered in them a movement of thought antithetical to that of the Romans. It was precisely the weakness of the early Christian movement that led to the deprecation of power and the glorification of peace and passivity – "turn the other cheek." And having no aspirations, as yet, to rule, they said, "Render unto Caesar the things that are Caesar's." The resentment of the early Christians was thus sublimated in a merely psychic rebellion, in which all the values of the Roman rulers were repudiated in countervalues.

Long before the rise of Christianity, however, the values of other high and mighty powers were also repudiated and inverted. The two cases in which Nietzsche is most interested are ancient Israel and Greece; and since he provides no extended analyses of those cases, we shall do so.

THE INVERSION OF VALUES IN ANCIENT ISRAEL

The Jews – a people "born of slavery," as Tacitus and the entire ancient world say, "the chosen people among the peoples," as they themselves say and believe – the Jews have accomplished that miraculous feat of an inversion of values, thanks to which life on earth has acquired a new and dangerous allurement for a couple of millennia: their prophets have combined "rich," "godless," "evil," "violent," and "sensual" into one and were the first to use the word "world" pejoratively [*zum Schandwort gemünzt*]. In this inversion of values (which includes rendering the word "poor" as synonymous with "holy" and "friend") lies the significance of the Jewish people: with them the slave-rebellion in morals begins. (*Beyond Good and Evil*, 195)

In an historical approach to the biblical narratives concerning the bondage and the Exodus, we need to begin with Egypt, the scene of the drama. The internal crisis suffered by Egypt during the reign of Amenhotep IV (the Amarna period) contributed substantially to the weakening of the empire in Syria and Palestine. Not too long afterwards, however, Rameses I (1319–1318 BC) and especially Sethi I (1318–1299) actively pursued a policy of regaining the Asiatic possessions. In accordance with that policy the capital of the empire was moved from the traditional metropolises – Thebes in the south and Memphis in the north – into the eastern delta at the gateway that led to Palestine and Syria. The new site was called Rameses.

This name occurs in the opening paragraphs of the Exodus narratives where we are told that the children of Israel were engaged in forced labor, building the Pharaoh's store-cities Pithom and Rameses (Ex. 1:11). As this is one of the few Egyptian place-names in the narratives, it provides a clue to the dating and localization of the Israelite stay in Egypt. It was Rameses II (1299–1232) who had built the new capital. The Egyptian court poets of the era are in agreement with the biblical writers that the great city lying between Palestine and Egypt was "filled with food and provisions."[1] If the city was named after him and the Israelites participated in its construction, it follows that he was the Pharaoh of the period of their bondage. It is also likely that either he or his son Merneptah was the Pharaoh of the exodus.

Favoring the latter possibility is the famous Merneptah stele in which he boasts of victory over several Asiatic peoples. On that stele we meet with the one and only instance of the name "Israel" in ancient Egyptian texts: "Israel is laid waste, his seed is not."[2] This, however, is not to be taken literally since Egyptologists inform us that destroying the "seed" of one's enemy was a conventional boast in that period. During Merneptah's reign Egypt was in fact troubled by difficult wars in Libya and Palestine, and Merneptah had to put down several rebellions, as we learn from his victory hymn. Perhaps the unrest in the empire and the troubles of the regime created an opportunity for the Israelites and other Semitic groups to throw off their yoke and to leave Egypt. There is good reason to believe that there existed at the time a great mass of Semites in the delta area, a "mixed multitude" in the words of the Scriptures (Ex. 12:38), who felt akin to the Israelites. That is the ground, perhaps, for Pharaoh's anxious observation that the people of Israel "are too many and too mighty for us" (Ex. 1:9). This mass, he fears, will join Egypt's enemies in any future wars that should befall her. Following Pharaoh's intensification of the oppression, the task of leading the oppressed out of bondage fell upon the shoulders of a great historical personality – a "higher type", in Nietzsche's terms – whom the Bible calls Moses.

There are, then, features of the biblical narrative that cannot have been invented out of whole cloth. Take the name "Moses" itself. It is a common Egyptian name-element as in "Thutmose" or "Thutmosis." Many ancient Egyptian names are based on the combination of a god's name with the root *msy*, meaning "to give birth to." Accordingly, such a name might be translated as "the god so-and-so is born" or "the god so-and-so has given birth (to him)." In the case of the name "Moses," the god's name is absent and only the element *msy* remains. That is not unique, however, since there is occasional Egyptian evidence for this sort of short name.[3]

From this it would not be extravagant to infer that the individual called "Moses" in the narratives received his name in Egypt. As the son of Semitic parents he, or they, could have been in Egyptian service. That the children of such persons were frequently given Egyptian names is confirmed in documents from as early as the Middle Kingdom. This might have been the case for Moses. But it is highly unlikely that he was an Egyptian, as was once fashionable to believe. He is described as having stood up for his "brothers" by smiting an Egyptian who was beating a Hebrew worker. When the killing became public knowledge, Moses fled the country, which suggests that he was not a member of the royal family, but rather subject to Egyptian law and justice.

There is, then, nothing inherently implausible in the historical role attributed to Moses. Seeing his people weighed down by forced labor, and being himself faced with a desperate situation, he takes the initiative and organizes and leads their departure. He has learned something about the Egyptian government and royalty from close quarters; and he knows the guarded borders and the terrain beyond since he had once escaped past them. That he was a man of extraordinary talent and courage is evident throughout the narratives.

In the course of their sojournings in the wilderness, the Israelites and those who fled with them were welded into a comparatively unified organization. The process of unification probably began in Egypt itself, for otherwise it would be difficult to understand how a large mass of people were able to co-ordinate and organize their flight. We have to assume that they had already shared something cultural in common, such as a religious faith, in connection with which they had requested from the Pharaoh permission for a pilgrimage. Equally important, of course, is the fact that they shared a servile and ignominious status in the Egyptian state system. This imparted to them a sense of common fate which made it possible for Moses and his helpers to transform them into a people.

In sociological terms we may see several conditions combining to produce this result. In the Egyptian-Semitic dichotomy, the Semites were not only oppressed but also held in contempt. The Israelites, in turn, together with the "mixed multitude" that went out with them, greatly resented their oppressors and deeply despised the Egyptian *corvée* state. The sharp cultural antithesis must be taken into account. The servile status of the oppressed Semites served to accentuate the already existing socio-cultural differences between them and their over-lords. Thus the Israelites experienced *ressentiment* and in due course inverted the beliefs and values of their oppressors. Among the people this probably occurred in an inchoate psychological fashion. However, it was Moses who, with a firsthand knowledge of Egyptian ways, was able to make this *ressentiment* explicit in the new religious idea that repudiated not only Egyptian values but the polytheistic worldview in general. What enabled him to play this role was, on the one hand, the negation of the oppressor's culture, which was already implicit in the feelings and thoughts of the slave people, and, on the other, certain positive ethical precepts with respect to human relations – precepts which can be traced back to the patriarchs, Abraham, Isaac, and Jacob.

There can be no doubt that the ethic expressed in the decalogue and covenant, notably the striking concern for justice, has its roots in the slave experience of the people. A basic principle of the Mosaic legislation is the protection of the weak from the power of the strong – what Nietzsche calls "slave morality." This is derived from the Israelites' humble and lowly condition and their redemption. Moses taught that God (Yahweh) requires humility, which is the reverse of hubris. Other polytheistic beliefs were also rejected. In place of a pantheon subordinate to an impersonal supradivine force (similar to *moira* in Greek religion), there emerged the belief in one personal, transcendental God of Justice who comes to the aid of the nobodies.

This was the transvaluation which enabled Moses to accomplish his organizational mission. Faced with the task of integrating the Israelites with the "mixed multitude" and overcoming divisiveness, he brought them the idea that they were to become a united people in a covenant with Yahweh – an idea that was the antithesis of the Egyptian "house of bondage" and its worldview. From a sociological point of view, Moses' stress on morality and justice served to mend the divisions in the multitude and to create a people. Such an entity could only be created if its outstanding moral qualities overcame a divisiveness and enhanced solidarity.

If the Decalogue and the covenant are sociologically bound up with the creation of the Israelites as a people – a united tribal confederacy – then we should expect the laws in question to reflect that concern. The moral code would then be designed to discourage any form of conduct threatening to the solidarity of the confederacy; and not only of conduct but even of attitudes which might undermine the group's unity. And, indeed, we read in the decalogue, "Thou shalt not covet." The members of the newly-formed confederacy are enjoined not to envy, or to do anything else for that matter which might spoil the emerging communal life. In this light, this commandment and the others constitute a major means of forging the communal consciousness of the Israelites as a people. If Moses led out of Egypt a "mixed multitude" – a heterogeneous agglomeration of tribes and clans – then the moral injunctions we read in Exodus 20*f.* are exactly what he would have required for the task of transforming that mass into a unity.

A similar logic applies to the commandments on adultery and re-membering the sabbath. Few actions are as destructive of community as adultery and unbridled sexual rivalry. Remembering the sabbath day to keep it holy is an institution of great antiquity in Israel which goes back to the origins of Yahweh-worship and Moses' original inspiration. One finds the commandment in all the traditions of the Pentateuch; the sabbath is a day consecrated to Yahweh. The sabbath is so distinctively Israelite that its roots cannot be traced to other more ancient cultures. In Exodus 20: 1–17 Israel is commanded to observe the sabbath in order to remember its slavery and deliverance. Here, again, we see an institution that is entirely congruous with the conditions of the Mosaic era. A formerly slave people creates a day of rest from their labors – for themselves, for their servants, for the *gerim* (strangers), and even for the cattle. So again we may invoke the concept of *ressentiment* and suggest that at least in one of its aspects the Israelite sabbath emerged as a repudiation of Egyptian bondage.

A comparison of the ancient Mesopotamian Code of Hammurabi, discovered in 1901, with the Mosaic legislation also tends to support the interpretation offered here in the light of *ressentiment*. The Hammurabi Code explicitly takes account of social-class distinctions in the administration of justice:

> If a nobleman has destroyed the eye of a member of the aristocracy, they shall destroy his eye.
> If he has broken another nobleman's bone, they shall break his bone.
> If he has destroyed the eye of a commoner, or broken the bone of a commoner, he shall pay one *mina* of silver.
> If he has destroyed the eye of a nobleman's slave or broken the bone of a nobleman's slave, he shall pay one half his value.
> If a nobleman has knocked out a tooth of a nobleman of his own rank, they shall knock out his tooth.
> If he has knocked out a commoner's tooth, he shall pay one-third of a *mina* of silver.
> If a nobleman has struck another nobleman's daughter and has caused her to have a miscarriage, he shall pay ten shekels of silver for her fetus.
> If that woman has died, they shall put his daughter to death.
> If by a blow he has caused a commoner's daughter to have a miscarriage, he shall pay five shekels of silver.
> If that woman has died, he shall pay one-half of a *mina* of silver.[4]

The relevant comparison in the Mosaic legislation is the famous "eye for eye, tooth for tooth" passage (Ex. 21:22–5):

if men strive together, and hurt a woman with child, so that her fruit depart, and yet no harm follow [to her], he shall surely be fined, according as the woman's husband shall lay upon him; and he shall pay as the judges determined. But if any harm follow, then thou shall give life for life, eye for eye, tooth for tooth, hand for hand, foot for foot, burning for burning, wound for wound, stripe for stripe.

This is then followed by a passage that points directly to the fundamental difference between the two codes:

And if a man smite the eye of his bondman, or the eye of his bondwoman, and destroy it, he shall let him go for his eye's sake. And if he smite out his bondman's tooth, or his bondwoman's tooth, he shall let him [*sic*] go free for his tooth's sake. (Ex. 21:26–7)

Clearly, the Mosaic code strives for a more egalitarian and universal adaptation of the punishment to the offense. The stratified social structure of Mesopotamia under Hammurabi is reflected in his system of laws, just as the democratic-egalitarian tribal confederacy is reflected in the Mosaic laws, which strive to protect several categories of individuals – strangers, widows, orphans, and even thieves (Ex. 22:20*f*) – from physical injury, oppression and various indignities. The injured slave is freed not because he is damaged property but because he is an oppressed human being. It is for that reason that the loss of a tooth no less than the loss of an eye represents abuse.

The concern for both the individual and the community is evident in other laws as well. Not only is lending at interest outlawed, but a creditor who has taken a poor man's garment as a pledge must return it every night (Ex. 22:25). One may not take advantage of the poor. Yet justice requires that one favor neither the poor nor the rich (Lev. 19:15*f*). In the law of the sabbatical year the social motivation comes to the fore. The land is to be worked for six years, but in the seventh it is to rest and lie fallow "so that the poor of the people may eat; and what they leave the beast of the field shall eat" (Ex. 23:10*f*). There is, then, throughout these ordinances an expression of genuine concern for underling and creatures alike. The motivation behind these laws is stated again and again: Israel must not oppress for she was an oppressed stranger in the land of Egypt (Ex. 22:20).

We see in this brief sketch the *sociological* fruitfulness of *ressentiment* and its role in ancient Israel's inversion of the noble values. And we see that it is historically accurate to describe the Mosaic code of justice as a "slave morality," inasmuch as it emerged out of Israel's slave experience. But, as we remarked earlier, sociological usefulness is not the same as philosophical validity. The fact that *ressentiment* is a useful analytical tool does not mean that the ethical values taught by Moses and the later Hebrew prophets can be dismissed, denigrated, or relativized just because they arose out of a slave experience. Nor can those values be regarded as necessarily "decadent" because of their origin in a slave experience. If Nietzsche or any other philosopher wishes to reject those values in favor of others, he has to give good reasons for doing so, which Nietzsche nowhere does. Let us confront him for a moment with the contrast we have just presented between the Hammurabi Code and the laws of Moses. The former, evidently, was the code of a society in which the nobility either ruled or possessed superior power. Would Nietzsche prefer the Hammurabi Code because it is "noble" and reject the Mosaic code because it is a slave morality? There are two lengthy passages which may help us to understand Nietzsche's attitude in this regard:

In the Jewish "Old Testament," the book of divine justice, there are human beings, things and speeches in so grand a style that the Greek and Indian literatures have nothing to compare with it. One stands with terror and reverence before these mighty remnants of what humanity once was, and will have sorrowful thoughts about ancient Asia and its protruding little peninsula Europe, which wants in all respects to signify as against Asia the "progress of man." Admittedly, whoever is himself merely a weak, tame domesticated animal, knowing only the needs of domestic animals (like our educated people of today, including the Christians of "educated" Christianity), has no cause for astonishment or sorrow among these ruins. The taste for the Old Testament is a touchstone in the consideration of "great" and "small." Perhaps he [the tame, domesticated European] will find the *New* Testament, the book of grace, rather more after his heart (it is full of the real, tender, musty, fanatic [*Betbrüder-*] and small-soul smell). To have attached this *New Testament*, a form of rococo of taste in every respect, to the *Old Testament* to create *one* book, as the "Bible," as "the Book *par excellence*," that is perhaps the greatest presumption and "sin against the spirit," that literary Europe has on its conscience. (*Beyond Good and Evil*, 52)

And in the *Genealogy* Nietzsche wrote:

I do not like the "*New Testament*," as should be plain. I find it almost disturbing that in my taste concerning this most highly esteemed and over-rated work, I should stand alone. (The taste of two millennia is against me.) But it can't be helped! "Here I stand, I cannot do otherwise." I have the courage of my bad taste. The *Old* Testament, that is something different: all honor to the Old Testament! There I find great human beings, a heroic landscape, and something that is rarest in the world, the incomparable naïveté of the *strong heart*; furthermore, I find a people. In the New one, in contrast, I find nothing but petty sectarianism, mere rococo of the soul, mere tortuous phrases, nooks, queer things, the air of the conventicle, not to forget an occasional whiff of bucolic sweetness, which belongs to the epoch (*and* to the Roman province) and which is not so much Jewish as Hellenistic. (*On the Genealogy of Morals*, III, 22)

We see, then, that Nietzsche had a very positive view of the Old Testament. In assessing these statements, however, we have to recall that for Nietzsche elements of the "master" and "slave" moralities are fused in reality. So what Nietzsche appears to be saying in the passages just quoted, is that although the Old Testament is largely the product of a slave revolt in morals, the Israelite inversion of values, as embodied in the Old Testament, has nevertheless preserved noble elements: "great human beings, a heroic landscape," "the incomparable naïveté of the strong heart," and "people, things and speeches in an incomparably grand style." In contrast, Nietzsche views the New Testament as so thoroughly a product of the slave morality as to be devoid of noble elements. Let us note as well, however, how often *taste* is stressed in these passages: "The taste of two millennia is against me." "I have the courage of my bad taste." "The taste for the Old Testament is a touch-stone in the consideration of 'great' and 'small'." The New Testament is a "form of rococo of taste in every respect." We must take Nietzsche at his word. His preference for the Old Testament is essentially a matter of taste; it is an aesthetic judgment having nothing to do, ostensibly, with the moral content of the Old and New Testaments.

NOTES

1 George Steindorff and Keith C. Seele, *When Egypt Ruled the East* (University of Chicago Press, Chicago, 1971), p. 257.
2 James B. Pritchard, *The Ancient Near East* (Princeton University Press, Princeton, 1958), vol. I, p. 231.
3 J. W. Griffiths, "The Egyptian Derivation of the Name Moses," *Journal of Near Eastern Studies*, 12 (1953), pp. 225–31.
4 Pritchard, op. cit., pp. 161–2.

5

The Inversion of Values in Ancient Greece

The *Iliad* and the *Odyssey* depict a society that takes pleasure in eating and drinking, in wealth and power, in skill in archery and shipbuilding, in the many details of pastoral life and in all the natural sights of the Greek world. It was a society conscious of its strength and success, and eager to hear itself praised. It was a society, in a word, that Nietzsche very much admired. By the end of the eighth century BC, however, the type of primitive monarchy pictured in the Homeric epics was no more. Although there remained hereditary kings, their role was reduced to that of magistrates with little power. The gradual breakdown of the primitive monarchies turned to the advantage of the powerful noble chiefs who became the new masters of the city-state. As we have seen, they took immense pride in their noble blood, defining themselves as the "good," "the best men," "the great and good," and so on. As large landowners and warriors they devoted themselves to raising horses. In battle the noble knight approached his adversary clad in heavy brass armor, his head enclosed in a visored helmet and his body protected by metal or leather sheets. He held a shield in his left hand and a long lance in his right, while a sword hung by his side. The war-horse, whether yoked to a chariot or mounted, was the distinctive mark of these aristocrats who were used to taking the lion's share of the booty acquired through raids on land and piracy at sea.[1]

In the seventh century BC, however, the expansion of commerce, and with it mining, manufacturing, and shipbuilding, brought about significant social and political changes. Money now increasingly replaced the old natural economy, and the large noble landowners were advantageously placed to gain most from these developments. They owned and controlled the fields, forests, vineyards, olive groves, mines, and quarries. They built large, slave-manned galleys and made forays into foreign lands and, through trade and outright piracy, returned with wealth in many forms. In this way the noble aristocracy slowly changed its character. Fluid capital in the form of precious metals, and not only landed wealth, now gave the nobles their power.

The nobility, however, was not the only class to benefit from the enormous volume of trade reaching from one end of the Mediterranean to the other. In most city-states the *demiourgoi*, the artisans and traders, were able to take a substantial portion of the profits and thus to form an intermediary class between the nobility and the *thetes*, the hired laborers. Although the new commercial class owned neither land nor horses, they had gained sufficient wealth to arm themselves and to form a heavy infantry called "*hoplites*." In time they became strong enough to threaten the position of the knights and even to defeat them in battle. A hybrid aristocracy now emerged in which the noble-warrior landowners retained their social honor, but in which wealth, whatever its source, rose in the scale of social values. The old aristocracy was becoming a plutocracy.

In the many city-states affected by these developments, power was always concentrated in the hands of the few, the *oligoi*, which is why the Greeks called this form of government "oligarchy." The meaning of this term for the Greeks was very different from that of "aristocracy", a concept which had earlier referred to the "best men," the ancient nobility, but which the future philosophers, Plato and Aristotle, would reserve for those who were morally and intellectually the best. Oligarchy always meant the division of the populace into two classes, only one of which had the right to participate in government. Extreme oligarchy was a situation in which the supreme magistrate commanded such great wealth and ruled over so many subjects that for all practical purposes sovereignty rested in the hands of one man. The main defect of the oligarchic system was that it created more and more inequality, even among the privileged. The monopolistic control of the magistracies gave to a few families and clans such great power that the majority of the economically privileged, who were excluded from government, refused to submit to this state of affairs. Aristotle informs us that time and again oligarchs were assailed: ". . . the nobles frequently form parties among the common people and among their friends, and so bring about a suspension of government, and form factions and engage in war with one another" (*Politics*, II, VII, 6–8).[2]

While the oligarchic rulers monopolized power in the cities, the less-skilled artisans struggled to make a living and the unskilled *thetes* were often destitute. Judging from Hesiod's description of the lot of the peasants, their position was steadily worsening. The small patches of land that scarcely sufficed for scratching out the necessities of life were increasingly swallowed up by the giant estates. The lands of the nobles, on the other hand, were expanding, since they were protected from permanent alienation by the kinsman's right to buy back any portion of the original inheritance which may have been sold. Furthermore, the nobles had successfully encroached on the traditionally communal pasture grounds. So although there was a middle stratum of cultivators who possessed their yoke of oxen for ploughing and who were capable in times of war of arming themselves at their own expense, the majority of the agricultural population lived in privation. In bad years, when the harvests were inadequate for both subsistence and sowing, they were compelled to borrow grain from the neighboring lord and to return it with substantial interest. Insolvent debtors, together with their wives and children, thus fell into the hands of the creditors. What made the lot of the impoverished peasants utterly hopeless was the fact that the grasping noble lords were in a position to bribe the magistrates, who were also members of the powerful and privileged circles.

Hesiod, who may have been a contemporary of Homer's, and whose *Works and Days* may go back to the eighth century BC, was both a witness and a victim of those "crooked" verdicts. He had come from Ionia to the mainland, settling in Boeotia where conditions were hard. He belonged to the class of small peasants and thought little of the nobles for whom Homer had sung. For Hesiod, whose primary concern is the daily struggle for existence, the nobles were not "sons of Zeus," but "devourers of the people." *Works and Days*, in one of its aspects, is a practical handbook written for Hesiod's brother, Perses, who is an ineffective manager and thus in need of advice about farming. But the other aspect of *Works and Days* is more important for our purposes. It is the earliest Greek text in which one can discern the phenomenon of *ressentiment* and the beginning of the inversion of values. Hesiod compares the unhappy wretches who have fallen into the clutches of the oppressors with the plight of a nightingale caught in the talons of a hawk:

> And now, for lords who understand, I'll tell a fable: once a hawk, high in the clouds, clutched in his claws a speckled nightingale. She, pierced by those hooked claws, cried, "Pity me!" But he made scornful answer: "Silly thing. Why do you cry? Your master holds you fast, you'll go where I decide, although you have a minstrel's lovely voice, and if I choose, I'll have you for a meal, or let you go. Only a fool will match himself against a stronger party, for he'll only lose, and be disgraced as well as beaten." Thus spoke the swift-flying hawk, the long-winged bird.

Hesiod then proceeds to assure Perses that *justice* will win out in the end.

> O Perses, follow right; control your pride. For pride is evil in a common man. Even a noble finds it hard to bear; it weighs him down and leads him to disgrace. The road to justice is the better way, for justice in the end will win the race and pride will lose: the simpleton must learn this fact through suffering. The god of Oaths runs faster than a crooked verdict; when justice is dragged out of the way by men who judge dishonestly and swallow bribes, a struggling sound is heard; then she returns to the city and the homes of men, wrapped in a mist and weeping, and she brings harm to the crooked men who drove her out. But when the judges of a town are fair to foreigner and citizen alike, their city prospers and her people bloom; since peace is in the land, her children thrive; Zeus never marks them out for cruel war. Famine and blight do not beset the just, who till their well-worked fields and feast. . . .

> The deathless gods are never far away; they mark the crooked judges who grind down their fellow-men and do not fear the gods. Three times ten thousand watchers-over-men, immortal, roam the fertile earth for Zeus. Clothed in a mist, they visit every land and keep a watch on law-suits and on crimes. One of them is the virgin, born of Zeus, Justice, revered by all the Olympian gods. . . . Beware, you lords who swallow bribes; try to judge uprightly, clear your minds of crookedness. *He hurts himself who hurts another man, and evil planning harms the planner most. . . .*

The son of Cronos made this law for men: that animals and fish and winged birds should eat each other, for they have no law. But mankind has the law of Right from him, which is the better way. . . .

I say important things for you to hear, O foolish Perses: Badness can be caught in great abundance, easily; the road to her is level, and she lives nearby. But Good is harder, for the gods have placed in front of her much sweat; the road is steep and long and rocky at the first, but when you reach the top, she is not hard to find.

That man is best who reasons for himself, considering the future. Also good is he who takes another's good advice. But he who neither thinks himself nor learns from others, is a failure as a man.[3]

It is from Hesiod that we first hear the negation of the prevailing "master morality." He is the first Greek writer to have repudiated the values of the high and mighty, replacing them with countervalues which Nietzsche has labeled "slave morality." Hesiod calls for *justice*, and his conception of what is just and unjust converges with that of the Israelite inversion of values and, in particular, with the conception of justice held by the Hebrew prophets of the eighth century BC, and to be considered later. Hesiod converges in still other respects; he calls for and yearns for *peace*, and he affirms that the social interdependence of human beings is such that one cannot avoid harming oneself when one harms another, which suggests the Golden Rule. He rejects analogies from the animal kingdom – in which one species preys upon another – as rationales for predatory, oppressive and exploitative behavior among humans; and he underscores that *law*, operating in accordance with Right and Good, ought to replace the naked exercise of power. Finally, he explicitly introduces Reason as a virtue.

Indeed, virtually all the critics and opponents of oligarchy were united on this point: publication of the laws was necessary. The people had had enough of the "crooked" sentences handed down by the nobles as the presumed will of the gods, but which were only too often a cynical exploitation of their power. Many generations had waited in vain for the judges, delivering sentence under solemn oath, to change their ways, but patience had finally run out and the people now demanded that the laws be written down.

The tension and strife in the countryside between the landlords and the impoverished peasants gave rise to a new temper and new laws. About 600 BC, Delphi broke away from the tribe of the Phocians and became a highly influential religio-moral institution in Greek life. The oracle of Delphi was famous. Its priests had an old tradition of their god, Apollo, as the originator of purification from the guilt of blood. They added to that tradition, and soon made him the interpreter of Greek ethics and the founding advocate of Greek law. The gist of the oracle's ethical teaching was the need for moderation, the need to remember that a boundary is set to all things, which they must not overpass. The lust for gain and power which had provoked the bloody conflicts must come under the regimen of ethical rules. "Nothing in excess" must become the principle governing social life.

After Hesiod and the new Delphic ethical doctrine, a new stage in the Greek inversion of values is met with in the work of the great lawgiver Solon, whose aim was to apply the principles of limit and moderation to the spheres of economic and political life. Solon sought to introduce, into a society torn by bloody strife between rich and poor, the ideal of social equality. He strove to prevent the mighty from exerting the power of their wealth without limit, and he also tried to better the conditions of the poor. He cancelled the debts accumulated on mortgage by the poor peasantry, and he fixed a limit to the size of landed properties. At the same time he restricted, by means of sumptuary laws, the right of the wealthy to flaunt their wealth. He attempted, in addition, to re-establish the peasants as freeholders on their land; and by opening Attica to foreigners who practiced some skilled craft, he fostered the rise of manufacturing; this was to prove, in the long run, a great boon to the poor by rescuing them from the dependence and misery of a purely agricultural way of life. To protect the weak and the needy he allowed any Athenian citizen to undertake, on another's behalf, a prosecution for a criminal offence. In a word, it was Solon's purpose, as we learn from his elegiac poems, to institute a general rule of balanced equality (*isonomy*), in which no class could claim either social superiority or political privilege:

> I gave to the people as much esteem as is sufficient for them, not detract-
> ing from their honor or reaching out to take it; and to those who had
> power and were admired for their wealth I declared that they should have
> nothing unseemly. I stood holding my mighty shield against both, and did
> not allow either to win an unjust victory.[4]

In his political reforms, Solon enacted a set of rules as instructions for the officials of the state, intended to control their administrative action. Treating the officials as servants of the law, he promulgated the law in writing so that a written code superseded the unwritten tradition. He thus founded a constitutional scheme in which the law was sovereign; and in order to ensure that the officials would conduct themselves as servants of the law, Solon made them responsible to a *public* court, the institution of which was his great innovation. This was a popular court consisting of thousands of judges throughout Attica, in which the poorest of citizens could sit and judge and which had the authority to review the conduct of every official at the end of his term of office. In effect, then, as Aristotle observed, Solon's reforms implicitly laid the foundations for Athenian democracy.[5]

One must not suppose, however, that either Solon or later legislators were able to do their work free from the raging passions of civil conflict. The mandate of the legislator was precisely to put an end to the bloodshed by mediating between the two sides and bringing about a reconciliation. To accomplish this he was invested with extraordinary powers and thus became the supreme head of the city. The legislator's mission was considered temporary, either for an indeterminate period until the task was accomplished or for a fixed period of one, five, or even ten years. In all cases it placed supreme authority in the hands of a single man. Aristotle called it an "elective tyranny."[6] Once the society had been saved by the efforts of the "elected tyrant," he returned to private life.

Not surprisingly, the nobility and the wealthy did not always submit to the legislator and accept his compromise. In such cases the people had recourse to an extreme expedient; they placed themselves in the hands of a tyrant, hoping thereby to gain some material improvement of their lot. The term "tyrant" probably originated in the East and signified master or king, as did its equivalent, *basileus*. On account of its origin and its association with the despots of the East, it was applied by the opponents of tyranny in a derogatory sense to those who had gained absolute power by insurrection rather than lawfully. An aspiring tyrant typically began his career as a demagogue leading the discontented masses against the nobles and other wealthy elements. The multitude followed him blindly so long as he promoted its welfare.

Tyranny did not establish itself throughout Greece. It emerged, evidently, in those cities where the commercial and manufacturing interests had created a markedly oppressive regime and where, therefore, a persuasive demagogue could mobilize the masses for an assault upon the privileged classes. Tyranny, as Thucydides observed, was associated with the growing importance of money.[7] The tyrants who made themselves the champions of the lower classes were generally restless, ambitious, and discontented members of the privileged camp. By virtue of the fact that they already held an office of state or a military command, the aspiring tyrants could mobilize a sizeable following of armed partisans and succeed in ousting the reigning monarch. While some of the tyrants were exceptionally ruthless in disposing of their actual and potential enemies,[8] all of them were preoccupied with controlling the lower classes by improving their economic conditions.

The urgent agrarian problem of restoring land to the impoverished and indebted peasants was often solved by distributing to them the property of the banished nobles. Solon had rejected the peasants' demand for a radical redistribution of property, but we no longer hear such a demand after the rule of the tyrant, Pisistratus. The most urgent task of all, however, was to find a means of appeasing the laboring masses of the city and maintaining the social peace. This the tyrants sought to accomplish by ensuring to the *thetes* a higher remuneration for their labor and a greater public respect. The tyrants also initiated and sponsored large-scale public works, aqueducts, breakwaters and monuments to the gods. This accomplished several things at once. It enhanced the tyrant's prestige; it kept the laborers occupied and less inclined to rebel; and it inspired in the people a civic pride which made them forget their loss of liberty. It is noteworthy, however, that tyranny nowhere endured for long. As a regime directed against oligarchy, tyranny persisted so long as it had the support of the people for whom it was a temporary expedient. Tyranny served the people as a weapon with which to demolish the citadel of the oligarchs, and once that was achieved, the people destroyed the weapon. The founding tyrant, placed on the pinnacle of power by the multitude, was almost always succeeded by a harsher and less capable ruler. As tyranny became less useful to the people, it also became more oppressive and thus doomed to death. And yet, it had fulfilled an historic role in activating the people and thereby contributing to the birth of Athenian democracy.

Prior to Solon's reforms, all loans to the peasants were made on the security of the person. The nobles ruled through nine archonships and the council of the Areopagus, the supreme tribunal. The Athenian government had become an oligarchy in which bloody conflicts between the notables and the masses had become the rule. Those were the circumstances in which Solon emerged as a great mediator and lawgiver. Both the rich and the poor, however, regretted Solon's appointment. The people had expected him to carry out a complete redistribution of the land, while the nobles had assumed that he would make only minute changes. Solon was opposed to the extremes of both parties, and though he was in a position to align himself with either one and become a tyrant, he followed his principles and incurred the animosity of both. In the face of such hostility, Solon left his homeland to trade and travel in Egypt, announcing that he would stay away for ten years.

As Solon's reforms had failed to achieve their aim, it was not for long that Attica remained peaceful. There were three factions in conflict, each representing specific social strata and areas of the land. The nobility of the plain were led by the Philidae; the merchants and fishermen of the coast by the Alcmaeonidae; the small peasants of the mountains by the Pisistratidae. Pisistratus (*c*. 560 BC) won the day by seizing the tyranny which the lower classes had offered to Solon in vain. Pisistratus introduced far-reaching agrarian reforms. By confiscating and breaking up the largest estates of the nobles and distributing the land to the peasants, he created a vigorous class of smallholders that came to play a significant part in Athens' affairs. He encouraged maritime commerce through a foreign policy that sent merchants to Thrace for gold and to the Hellespont for corn. At the same time he catered to the rural and urban masses by means of festivals in honor of Dionysus, by theatrical productions, and by the construction of magnificent edifices. And since he revitalized Solon's political reforms, he fostered among the people valuable political experience in the sessions of the Assembly and in the courts. Pisistratus appears to have enjoyed the support of the various classes; for in the words of *The Athenian Constitution* he ". . . always maintained peace and saw that all was quiet. For that reason it was often said that the tyranny of Pisistratus was the age of Cronos; for afterwards, when his sons took over, the regime became much more cruel. Most important of all the things mentioned was his democratic and humane manner. . . . He had many supporters both among the notables and among the ordinary people: he won over the notables by his friendly dealings with them, and the people by his help for their private concerns, and he behaved honorably to both."[9]

Cleisthenes (*c.* 525/4) continued the work of Solon by giving a more decisive form to the emerging democracy of Athens. His major aim was to prevent the return of oligarchy. To accomplish that aim he recognized that it would be necessary to destroy the strong political organizations which the nobles had created for themselves in their four powerful Ionian tribes. Under Cleisthenes the people gained more control over their affairs. After the overthrow of the last tyrant, Isagoras, Cleisthenes distributed all the citizens through ten tribes instead of the old four, thus mixing families so that men of all social backgrounds would have a share in running the state. The decimal system was applied to the entire political and administrative organization of the city. The Boule, or supreme council, consisted of 500 members, 50 from each tribe. Since there were only nine archons, a secretary was added to them so that each of the ten tribes would be represented in the college of magistrates. The army was made up of ten regiments called *phylai*, each commanded by a *phylarch*. On all occasions citizens were organized in groups of ten; and, indeed, the decimal system – a logical criterion and thus contrary to all previous forms of social organization – became an integral feature of the democratic regime. Thus the Athenians of the fifth century lived in accordance with the civil laws of Solon and the political laws of Cleisthenes.

In less than 20 years after Cleisthenes' reforms, the new Athenian democracy was put to a severe test with the great invasion by the Medes and Persians around 480 BC. She stood the test and came out strengthened. In the course of the war, however, the Areopagus, whose members were drawn in a large measure from the noble and wealthy strata, had magnified its role in public life. In 462, the democratic party, under the leadership of Ephialtes, launched an attack upon this institution, which had become a stronghold of the aristocracy. As a result, the Areopagus retained only its religious functions, while all other powers were withdrawn from it and transferred to the Assembly of the people and the Boule. Described as a champion of the people and "uncorrupt and upright in political matters,"[10] Ephialtes was assassinated by a political enemy thus paying with his life for his dedication to the democratic cause. But he had as his lieutenant a man who was to become the most famous representative of Athenian democracy – Pericles, the son of Xanthippus and the great-nephew of Cleisthenes.

Pericles solved the problem of how to enable the common people, totally engaged in the earning of a livelihood, to participate in government. The administration of justice throughout the country required thousands of judges and other officials. Public affairs demanded the occasional presence of all the citizens in the Assembly, and the continual presence of more than a third of them. But many citizens earned scarcely enough for their subsistence; how could they forgo the income even of a few days? And if such working people were to be excluded from government, why would this be a democracy rather than a timocracy or oligarchy? It was Pericles' innovation to require the state to pay the citizens who, in order to serve as officials, had to abandon their occupations. With this innovation there was no longer any reason for limiting public service to the proprietary classes.

Thucydides in his *History of the Peloponnesian War* devotes a chapter to "Pericles' Funeral Oration," an address Pericles delivered at a public ceremony honoring the memory of those who had been the first to die in the war. Here are a few of the words Thucydides attributed to Pericles on that occasion:

> Our constitution is called a democracy because power is in the hands not of a minority but of the whole people. When it is a question of settling private disputes, everyone is equal before the law; when it is a question of putting one person before another in positions of public responsibility, what counts is not membership of a particular class, but the actual ability which the man possesses. No one, so long as he has it in him to be of service to the state, is kept in political obscurity because of poverty . . .
>
> We give our obedience to those whom we put in positions of authority, and we obey the laws themselves, *especially those which are for the protection of the oppressed* . . .
>
> Our love of what is beautiful does not lend to extravagance; *our love of the things of the mind does not make us soft*. We regard wealth as something to be properly used, rather than as something to boast about. As for poverty, no one need be ashamed to admit it: the real shame is in not taking practical measures to escape from it. Here each individual is interested not only in his own affairs but in the affairs of the state as well:

even those who are mostly occupied with their business are extremely well informed on general politics – that is a peculiarity of ours: we do not say that a man who takes no interest in politics is a man who minds his own business; we say that he has no business here at all . . . Taking everything together, then, I declare that our city is an education to Greece, and I declare that in my opinion each single one of our citizens, in all the manifold aspects of life, is able to show himself the rightful lord and owner of his own person, and to do this, moreover, with exceptional grace and exceptional versatility.[11]

We see, then, that the story of the inversion of values in ancient Greece is largely but not exclusively the story of the origin and development of Athenian democracy. For as we shall see, Socrates and Plato, though opposed to democracy, were active participants in the inversion of noble-warrior values.

The foregoing sketch, like the preceding one on ancient Israel, is a supplement to Nietzsche's own discussion. Both sketches are intended to provide a fuller substantive understanding of his conception of the inversion of values. If, for the moment, we confine our attention to the Greek case, it might be interesting to confront Nietzsche with a few questions and to imagine how he would answer them. What would his attitude be toward the so-called "slave morality" of Hesiod? As we have seen, Hesiod called for justice, peace and reason. Would Nietzsche have an unqualified admiration for the "master morality" against which Hesiod reacted? Nietzsche remarks in the *Genealogy* on Hesiod's dilemma when he created his succession of cultural epochs and labeled them gold, silver, and bronze. Recognizing both the glorious and the violent and terrible sides of the world depicted by Homer, Hesiod found no way to handle the contradiction except by dividing one epoch into two and placing them in sequence: first, the epoch of the heroes and demi-gods as they survived in the memories of the nobles who viewed themselves as the true descendants of those heroes, and then the bronze epoch, the way in which that same epoch appeared to the descendants of the downtrodden and enslaved, an epoch that was hard, cold, cruel, bloody and destructive (*Genealogy*, I, 11). If Nietzsche thus sees both sides of the reality dominated by a "master morality," does he reject outright Hesiod's repudiation of the dark, oppressive side? Let us agree that Hesiod's call for justice was in fact motivated by *ressentiment*. Does that insight necessarily render Hesiod's conception of justice invalid? "Crooked" judges, drawn from the ranks of the nobility, constituted an integral element of the "master morality" of Hesiod's era. Would Nietzsche merely shrug off Hesiod's denunciation of such practices as the spiritual revenge of the impotent? The main thrust of Nietzsche's philosophical thought suggests that his admiration for the master morality was also an endorsement of all its elements. Moreover, his contempt for the slave morality was such that he would have refused even to consider the proposition that Hesiod's understanding of justice – like that of Moses and the prophets – possessed a lasting, trans-historical validity.

Similar questions might be asked about the work of Solon and the later reformers. In a large measure their efforts were directed against the prevalent form of the "master morality." How would Nietzsche regard Solon's efforts at mediation between the classes? In an earlier discussion we observed that Nietzsche himself appears to have had a mediatory and conciliatory vision with regard to Europe:

> There are today perhaps ten to twenty million people among the various nations of Europe who no longer believe in God. Is it too much to ask that they give one another a sign? Once they have thus come to recognize one another, they will also have made themselves known to others – they will at once become a *power* in Europe and, happily, a power *between* the nations! Between the classes! Between poor and rich! Between rulers and subjects! Between the most unpeaceful and the most peaceful, peace-bringing people! (*Dawn*, 96)

These words indicate that Nietzsche might have approved of Solon's efforts. But the question remains: on what ground would he approve? If he denies moral grounds, as he does, his approval can only rest on his personal taste. But can the ideal of peace among nations and classes be merely a matter of taste?

And what about Pericles? We have seen that Nietzsche is contemptuous of democracy throughout. But Pericles stresses, after all, that the democratic way of the Athenians does not make them *soft*. He calls attention to the noble and heroic elements of Athenian democracy. "The Spartans," he said, from their earliest boyhood, are submitted to the most laborious training in courage; we pass our lives without all these restrictions, and yet are just as ready to face the same dangers as they are . . . There are certain advantages, I think, in our way of meeting danger voluntarily, with an easy mind, instead of with a laborious training, with natural rather than with state-induced courage. We do not have to spend our time practising to meet sufferings which are still in the future; and when they are actually upon us we show ourselves just as brave as these others who are always in strict training. This is one point in which, I think, out city deserves to be admired.[12]

Given these words and the interpenetration of "master" and "slave" elements in any real political culture, it is almost certain that Nietzsche would have to acknowledge the heroic and noble elements of Athenian democracy. Were there no such elements in the democracy of his own time? Would he however reserve his admiration only for those elements, contemptuously rejecting the others? Pericles and the defenders of democracy since his time have argued that finding a collection of "wise" and "noble" individuals and leaving the government to them would be undesirable even if it were possible, for there is no reason to believe that the One or the Few are always wiser or better than the Many. It is doubtful in the extreme that Nietzsche would take such an argument seriously, since he cares not at all about the "herd."

It is interesting that although the Platonic Socrates was a severe critic of democracy, he nevertheless opposed the proto-Nietzschean view of the time, that "might is right." But before we review that debate between Socrates and his opponents, we need to discuss the second and third essays in the *Genealogy*.

NOTES

1   This historical sketch is based on J. B. Bury and Russell Meiggs, *A History of Greece*, 4th edn (with revisions) (St. Martin's Press, New York, 1985).

2   Aristotle, *Politics*, tr. H. Rackham, Loeb Classical Library, vol. 21 (Harvard University Press, Cambridge, Mass., 1977).

3   Hesiod, *Work and Days*, tr. Dorothea Wender (Penguin Books, London, 1987), sections 195–322.

4   Aristotle, *The Athenian Constitution*, tr. P. J. Rhodes (Penguin Books, New York, 1987), section 12.
5   Ibid., sections 9 and 10.
6   Aristotle, *Politics*, op. cit., III, IX, 4–7.
7   Thucydides, *History of the Peloponnesian War*, tr. Rex Warner (Penguin Classics, New York, 1988), I, 13.
8   Herodotus, *The Histories*, tr. Aubrey de Sélincourt (Penguin Books, New York, 1972), V, 92e–g.
9   Aristotle, *The Athenian Constitution*, op. cit., section 16.
10   Ibid., section 25.
11   Thucydides, op. cit., II, 40–1.
12   Ibid., II, 39.

6

Guilt, Bad Conscience, and Ascetic Ideals

The second and third essays in the *Genealogy* are concerned with the social origin of guilt and asceticism. For Nietzsche, the origin of guilt can be traced to an old and deeply rooted notion that every offense or injury, whether against an individual or a community, has its equivalent which can be repaid through the pain of the offender. The idea of equivalence between injury and pain emerged quite early in history as individuals entered into contractual relations as buyers and sellers, and especially as creditors and debtors. When someone borrowed money, or anything else for that matter, he had, of course, to promise that he would repay it. In order to inspire trust in his promise and to provide a guarantee of its seriousness and, indeed, in order to impress upon his own conscience the duty to repay, the debtor formed a contract with the creditor pledging that if he failed to repay the loan, he would substitute for it something else over which he had control – his freedom, his wife or children, even his life. We learn from history that when a debtor failed to repay, that gave the creditor the right, under the terms of the contract, to inflict upon the body of the debtor every imaginable indignity and torment – for example, cutting from it as much as appeared to be the equivalent of the debt. From earliest times there are documents of a legal kind stipulating in considerable detail which limbs and parts of the body may be taken by the creditor as compensation for the debtor's failure to repay the original loan. Nietzsche cites the Twelve Tables of Rome in which it was a matter of indifference whether the creditor cut off more or less in such cases. Nietzsche's main aim in this essay is to bring into relief the psychology of this form of compensation: it gives the creditor a kind of *pleasure*, the pleasure of being allowed to exercise his power freely upon one who is powerless, of violating another human being for the sheer enjoyment it brings. Nietzsche advances the proposition that the lower the creditor stands in the social order, the greater will be his enjoyment of such a deed, since in so "punishing" the debtor, the creditor gains a taste of higher rank and participates, as it were, in a right of the masters. It is in the sphere of legal obligations that the moral concepts of "guilt," "conscience," and "duty" had their origin, an origin thoroughly soaked in blood.

Writing before the dawn of the twentieth century and in the comparatively civilized conditions of the nineteenth, Nietzsche remarks that modern man – that tame, domesticated animal – finds it difficult to comprehend the degree to which cruelty in earlier times constituted an essential ingredient of pleasure. It was not so long ago, he reminds us, that princely weddings and public festivals included, as a matter of course, executions, torturings, or perhaps an *auto-da-fé*; and no noble household was without creatures on whom superiors could freely vent their malice and cruelty. It is a hard saying, says Nietzsche, but an indisputably human, all-too-human, principle that witnessing others suffer does one good, and making others suffer does one even more good. This observation of Nietzsche's is, however, not intended as grist for the mills of his pessimistic contemporaries, since he believes that in the days when humanity was not yet ashamed of its cruelty, life was more cheerful. Pessimism, virtually unknown in those early, *evil* epochs of the human race, first makes its appearance as men become ashamed of their instincts and begin to repress them by means of a morbid moralization. While the modern age abhors suffering and views it as an argument against existence, the earlier, heroic age was unwilling to refrain from causing others to suffer, and saw in it a fascination and a delight of the highest order, a genuine inducement *to* life. Homer attests to the fact that the Greeks knew nothing more palatable to offer the gods for the enhancement of their pleasures than the recurring spectacle of human cruelty. Wasn't that the meaning, after all, of the Trojan wars and other such tragic terrors? For Nietzsche there can be no doubt that such terrors were intended as festive spectacles for the gods. He suggests that the idea of "free will" was invented to ensure the absolute spontaneity of man in good and evil, so that the gods' interest in human events could never be exhausted; a totally deterministic world would have made human actions predictable and, hence, boring and wearisome for the gods.

Pursuing the origin of guilt in the buyer-seller, creditor-debtor relationship, Nietzsche offers us a variation on the old theme of "social contract." What happens when an individual offends against his community, to which he owes so much? Converging again with the sociological theory of Emile Durkheim, Nietzsche underscores that such an individual is above all a breaker of his contract with the community as a whole from which he has derived protection, benefits, and comforts. The offender or lawbreaker is a debtor who has not only failed to pay his debts to society, but actually attacked the creditor, thereby inflicting a wound on the body politic. The community-creditor heals its wound by depriving the offender of all the benefits and advantages which it had previously bestowed upon him, thus reaffirming its integrity. Punishment, in its earliest, "primitive" forms, treats the offender as a hated, disarmed enemy who has lost all rights; it is thus the rights of war and victory, with all of their mercilessness and cruelty, that the community exercises against the offender. It is therefore the phenomenon of war, according to Nietzsche, that has provided humanity with all the forms that punishment has assumed in the course of history. As a community's power increases, however, it takes the individual's offenses less seriously, since they are no longer as dangerous and destructive to the whole as they were previously. The stronger and more confident the community, the more moderate is its penal law. The "creditor" becomes more humane to the extent that he has become richer; he measures his wealth by the amount of injury he can endure without suffering from it. So the early conception of justice looked upon an offense as a debt that had to be discharged, while the modern conception tends to excuse the offender and set him free. This Nietzsche calls the "self-overcoming of justice," which has given itself the beautiful name of *mercy* – the exclusive privilege of those who are the most powerful.

Nietzsche now raises an interesting question with regard to justice and mercy. He rejects the view that the origin of justice is to be sought in the phenomenon of *ressentiment*. He insists that wherever justice is practiced and maintained it is a stronger power seeking a means of putting an end to the raging of *ressentiment* among the weaker powers (II, 11). If we follow the logic of Nietzsche's "contract" metaphor, and ask what justice might mean in the light of that metaphor, it would seem that justice implies balance, symmetry, reciprocity and even equality between the contracting parties. And we have seen in the case of ancient Greece that the "contractual relationship" between the noble lords and the peasants was one between social unequals, where one dominated and exploited while the other served and suffered. It was this state of affairs that engendered the *ressentiment* of Hesiod, whose call for justice was a call to protect the weak against the strong and to abolish the crooked practices of the powerful nobles who behaved in accordance with the motto "might is right." Clearly Hesiod was speaking for the relatively power-less, and it required the power of a Solon and the reformers who followed him, to try to establish some modicum of justice. So Nietzsche's point is that although the *demand* for justice emerged in the sphere of *ressentiment*, the weak and oppressed learned the concept of justice by observing the balanced, symmetrical and equitable relations among their noble masters. Nevertheless, if the principle of justice was merely implicit among the masters, it seems clear that it was first enunciated as an explicit principle in the ranks of the weak as something to be demanded from the strong.

Another interesting issue raised in these essays is the nature of human nature and conduct. Nietzsche dislikes the concept of "adaptation" for its passive connotation, for erroneously implying that human conduct is a "mere reactivity." This view, he maintains, misses the essence of life, its will to power. The notion of "adaptation" overlooks the ". . . essential priority of the spontaneous, aggressive, expansive, form-giving forces that impart new interpretations and directions . . ." (II, 12). In this respect Nietzsche anticipates and converges with the outstanding American founders of the Pragmatist movement in philosophy. William James brought to the fore the "insurgent character" of the organism as did John Dewey, who remarked: "it is absurd to ask what induces a man to activity generally speaking. He is an active being and that is all there is to be said on that score."[1] Similarly, George Herbert Mead stressed that the "I" represents freedom, spontaneity, novelty and initiative and that all these qualities are rooted in the human being's biological nature. Anticipating Freud as well, Nietzsche proposes that the "bad conscience" came into being as man found himself "enclosed within the walls of society and peace" (II, 16). The constraints that society imposes on the human instincts prevents them from discharging themselves outwardly and naturally. The instincts therefore turn inward, which Nietzsche calls "internalization." As human beings become civilized, they lose their former guides to healthy conduct, their unconscious and infallible regulating drives. With the civilizing process, says Nietzsche, humans were reduced to ". . . thinking, deducing, calculating, connecting cause and effect, these unfortunate creatures; they were reduced to their 'consciousness,' their weakest and most fallible organ!" (II, 16). Nietzsche's view of the relationship of the biological to the social side of the human being is, however, problematic. The first problem lies in his conception of thinking and consciousness, which is quite the opposite of what Darwin's research had led him to conclude.

For Darwin, the hominids' exodus from the trees and the permanent descent to the ground presented them with a host of new problems and challenges. They now had to cope with radically new circumstances and to develop new cognitive capacities. Accordingly, those individuals with larger and heavier brains adapted themselves more easily. The growth of the brain affected the shape of the skull, while the increased weight of the skull influenced the development of the supporting spinal column. Thus the earliest humans diverged from their apelike ancestors. Darwin calls our attention to a paradox: although humans have become the most powerful beings on earth, they entered the world as the weakest of nature's creatures. The human being comes into the world without great teeth and sharp claws for defense, and his physical strength and speed are by no means the greatest. His biological equipment, in and of itself, would prove quite ineffectual in the struggle for survival. It is therefore clear that the "secret" of human dominance lies elsewhere. It was the unique mode of interacting with nature that enabled humans to become dominant despite their natural weakness. In Darwin's words:

> . . . we cannot say whether man has become larger and stronger, or smaller and weaker, in comparison with his progenitors. We should, however, bear in mind that an animal possessing great size, strength, and ferocity, and which, like the gorilla, could defend itself from all enemies, would probably, though not necessarily, have failed to become social; and this

would most effectually have checked the acquirement by man of his higher mental qualities . . . Hence it might have been an immense advantage to man to have sprung from some comparatively weak creature.

The slight corporeal strength of man, his little speed, his want of natural weapons, etc., are more than counterbalanced, firstly by his intellectual powers, through which he has, whilst still remaining in a barbarous state, formed for himself weapons, tools, etc., and secondly by his social qualities which lead him to give aid to his fellowmen and to receive it in return.[2]

We shall need to say more about Nietzsche's misunderstanding of Darwin in a later chapter.

There is another problem with Nietzsche's view of the human being's biological nature. Implicitly employing "state of nature" as a heuristic device, Nietzsche proposes that once humans have left that state and entered into a social organization, all their old instincts of freedom are inhibited, punished and "internalized" – turned against themselves. In Nietzsche's words:

Hostility, cruelty, pleasure in persecuting, in attacking, in change, in destruction – all this turned against the possessors of such instincts: *that* is the source of the "bad conscience." (II, 16)

The parallels with Freud's theory are obvious. Like Nietzsche, Freud says of the instinctual energy (the *id*) that it "knows no judgment of value: no good and evil, no morality."[3] Freud's concept, "repression," is pretty much what Nietzsche meant by "internalization," and Freud's "super-ego" is functionally equivalent to Nietzsche's "bad conscience." And yet Nietzsche's consistent use of bellicose metaphors with which to describe human drives can be misleading. That is to say, that there is an element in Freud's theory that is missing from Nietzsche's. That element remains operative even in Freud's late and more pessimistic thoughts on the subject:

The transformation of "bad" instincts is brought about by two factors working in the same direction, an internal and an external one. The internal factor consists in the influence exercised on the bad (let us say, the egoistic) instincts by erotism – that is, by the human need for love, taken in its widest sense. By the admixture of *erotic* components the egoistic instincts are transformed into *social* ones. We learn to value being loved as an advantage for which we are willing to sacrifice other advantages. The external factor is the force exercised by upbringing, which represents the claims of our cultural environment, and this is continued later by the direct pressure of that environment. Civilization has been attained through the renunciation of instinctual satisfaction, and it demands the same renunciation from each newcomer in turn. Throughout an individual's life there is a constant replacement of external by internal compulsion. The influences of civilization cause an ever-increasing transformation of egoistic trends into altruistic and social ones by an admixture of erotic elements.[4]

Nietzsche anticipates Freud by recognizing the reality of repression – that life in society inevitably exacts a high toll from the human being in the form of the reduction or renunciation of instinctual satisfaction. But whereas Freud recognizes that when accompanied by Eros repression brings with it certain redeeming social qualities, Nietzsche appears to have nothing but contempt for what he calls the "herd-values" of co-operation and altruism. Moreover, Freud's view of the "ennoblement" of instincts is quite different from that of Nietzsche's. "The transformation of instinct," wrote Freud,

on which our susceptibility to culture is based, may also be permanently or temporarily undone by the impacts of life. The influences of war are undoubtedly among the forces that can bring about such involution; so we need not deny susceptibility to culture to all who are at the present time [the First World War] behaving in an uncivilized way, and we may anticipate that the ennoblement of their instincts will be restored in more peaceful times.[5]

As one reads Nietzsche's encomium to the instincts, however, it appears that for him the "ennoblement" of the instincts would mean removing the constraints of the "herd" and allowing only such constraints as are self-imposed by the "higher specimens." This is a fair inference from the numerous passages in which he speaks with admiration of conquerors and masters, "violent in act and bearing." "I used the term 'state,'" wrote Nietzsche, what is meant by that is obvious – a pack of blond predatory beasts, a conqueror and master race [*Herren-Rasse*], which, organized for war and with the power to organize, unhesitatingly lays its dreadful claws upon a populace, perhaps far superior in numbers but still formless and nomadic . . . Their work [of the conquerors and masters] is an instinctive, form-giving creation; they are the most involuntary and unconscious of artists – wherever they appear something new soon emerges, a ruling structure that *lives* . . . They do not know the meaning of guilt, responsibility or consideration, these born organizers; they exemplify that frightful artists' egoism that has the look of bronze and shows itself justified to all eternity in its "work" . . . (II, 17)

Some scholars have pointed to passages of this kind to support their contention that, however inadvertently, Nietzsche had thus helped to prepare the ideological ground for the Nazis. The general character of Nietzsche's rhetoric and his explicit use of the term "master race" made it not too difficult for such scholars to press their point. Walter Kaufmann has worked hard to refute this view, arguing that Nietzsche looked to art and philosophy, and not to race, to raise some men above the rest of mankind. Nevertheless, even Kaufmann has to acknowledge that Nietzsche's doctrine ". . . was dynamite insofar as it insisted that the gulf between some men and others is more significant than that between man and animal."[6]

In his closing remarks on "guilt" and "bad conscience" Nietzsche asks whether he has been erecting an ideal or knocking it down. He fancies that he has done both, by knocking down the old "bad conscience" and erecting in its place a new one whose meaning is the opposite of the old. Human beings have for too long had an "evil eye" for their natural inclinations. So the effort must be made to link the bad conscience to all the *unnatural* inclinations, to feel guilt in the presence of all ideas and yearnings that run counter to what is instinctual, natural, and animal in humans. To whom should one turn, Nietzsche asks, with such hopes of restoring the noble, masterful type of man in whom the natural impulses are free? It will come as no surprise that Nietzsche's answer returns to the central motif of *Zarathustra*: to realize the new ideal, a different kind of human spirit will be required, a spirit "strengthened by war and victory, for whom conquest, adventure, danger, and pain have become needs" (II, 24). The men of the future will require *great health* if they are to succeed in redeeming humanity from the will to nothingness, that is to say, the belief in God which, for Nietzsche, is the real nihilism.

ON THE MEANING OF ASCETIC IDEALS

In this third essay in the *Genealogy*, Nietzsche wants to show that asceticism is part and parcel of that same repose in nothingness ("God") that has produced the anti-life phenomena of guilt and bad conscience. Life, alas, has had no meaning apart from the ascetic ideal. Take the philosopher who is totally committed to his calling. To achieve the optimum he strives to remove every obstacle from his path, to remove any and all distractions, including marriage. Thus most of the great philosophers – Heraclitus, Plato, Descartes, Spinoza, Leibniz, Kant – remained single. Socrates is the conspicuous exception, Nietzsche notes, but he married ironically. A married philosopher belongs in comedy, he says, alluding to Socrates' role in Aristophanes' *The Clouds*. Consider the virtues of the philosopher, his inclination towards doubt, denial, suspension of judgment, analysis, balance, neutrality, objectivity, impartiality. But it is not only the philosophers but also the rest of humanity who subject themselves to an ascetic life – an asceticism, ironically, which is at the same time sheer hubris and godlessness. Take the Western attitude toward Nature and the way we violate her with machines and the heedless inventiveness of technicians and engineers. Take, too, the attitude towards ourselves, and the way we permit our bodies and minds to be tampered with. Indeed, viewed from a distant star, the earth as a whole must appear to be a distinctively ascetic planet, for its human inhabitants take great pleasure in inflicting great pain upon themselves. How else explain the ascetic priest, who appears regularly in every age? These poor souls do not yet realize that an ascetic life is a self-contradiction, a life in which an unequaled *ressentiment* rules. Under the ascetic ideal the instinctually rooted will to power strives for mastery not over obstacles to life, but over life itself.

The ascetic attitude is no less self-defeating in the pursuit of knowledge. Seeing differently and wanting to see differently are a necessary discipline and preparation for "objectivity." But what Nietzsche means by that term is not seeing in a detached or disinterested way – which is a nonsensical absurdity – but the capacity to control one's Fors and Againsts and to set them aside so that one can employ a variety of perspectives in the service of knowledge (III, 12). Our active and interpretive capacities cannot but include affect. Seeing and knowing are always perspectival; "and the more affects we allow to express themselves about a thing, the *more* eyes, different eyes, we can employ to observe a thing, the more complete will our 'conception' of that thing, our 'objectivity' be" (III, 12).

The ascetic ideal has infected every area, and has created a sick animal – the sickest of animals. The sick represent the greatest danger to the healthy. It is the sick, those who are failures from the start, the weak and the inferior, who undermine life. And yet they presume to monopolize virtue, employing a variety of devious ways with which to tyrannize over the healthy. The sickly are the unfortunate men of *ressentiment*: physically and mentally defective, they seek revenge against the fortunate and strong. They have poisoned the consciences of the fortunate with their own misery. They want the fortunate, higher specimens to be ashamed of their good fortune and to consider it disgraceful to be fortunate in the face of so much misery. Proclaiming once again his aristocratic doctrine, Nietzsche declares: Away with this "inverted world!" He urges the higher specimens to grasp their task rightly: they ought not allow themselves to be degraded to the status of an instrument of the inferior. The tasks of the higher and the lower ought to be eternally separate. It is the higher whose right to exist is a thousand times greater, for they are the warranty for the future (III, 14).

The sick herd needs a shepherd and a savior and that is the historical mission of the ascetic priest, who deals effectively with the most hazardous of all explosives accumulating within the herd – *ressentiment*. Every sufferer in the herd seeks the cause of his suffering, some living thing whom he can blame for his suffering and upon whom he can vent his anger. The priest persuades the sufferer that he himself is to blame for his pain, thus altering the direction of *ressentiment*. The notion of "sin" is itself a manifestation of sickness. "Sinfulness" is not a fact but only a psychological interpretation of a fact – physiological depression. The religio-moral interpretation of that fact is not only no longer binding, but positively debilitating. Unlike the sickly and the weak, the strong and well-constituted individual digests his experiences, including his misdeeds, as he does the toughest morsels of his meals. The dark melancholy of the weak has nothing to do with so-called sinfulness but rather with the severe inhibition of their physiological appetites.

Such melancholy in the herd, Nietzsche observes, is also combatted by "mechanical activity." A mechanical regimen alleviates suffering at least for the duration of the regimen, so the ascetic priest preaches to the herd "the blessing of work." The priest preaches, in addition, that the supreme pleasure is giving pleasure, doing good to others. Nietzsche proposes that the priest, by preaching "love of one's neighbor," actually excites the strongest, most life-affirming drive, namely, the will to power. That drive is gratified by virtue of the fact that doing good always imparts a sense of superiority to the giver of aid. The ascetic priest understands very well that doing good, helping and being useful, are the most effective means of consoling the physiologically inhibited members of the herd, who otherwise would injure one another in obedience to the same basic instinct. Nietzsche points to the beginnings of Christianity in the Roman world, where there existed in the lowest strata of society all sorts of associations for the poor and the sick. Associations of mutual aid became the major antidote to depression among the lowly. The petty pleasure produced by mutual helpfulness was consciously employed by the early leaders of the Christian congregations. The formation of a herd – a community or congregation is itself a significant victory over depression. All the sick and weak naturally strive after a herd organization as a means of counteracting their feeling of weakness. The strong, in contrast, are naturally inclined to separate. Nietzsche is consistently contemptuous of mutual aid among humans, but admiring of the "strong" whose lust for power he describes not in spiritual (cultural) but in political terms. Every oligarchy conceals the lust for tyranny, each of its members trembling with the tension he feels in maintaining control of that lust. "So it was in Greece, for instance," says Nietzsche, "and Plato attests to it in a hundred passages, and he knew his own kind – *and* himself" (III, 18).

For Nietzsche, guilt, the bad conscience, and the overarching ascetic ideal are primarily the product of Christian morality. Far from having improved humanity, he contends, Christian "medications" have merely tamed, emasculated and discouraged humanity, thus actually harming it.

The ascetic priest has ruined humanity's psychical health wherever he has come to power. This is the context in which Nietzsche coolly states, "I do not like the 'New Testament,' that should be plain . . ." (III, 22). Why? Precisely because it is permeated throughout by the values of the ascetic ideal. It lacks any trace of good breeding, as is evident in the fuss those pious little men make over petty vices (III, 22). Under Christian influence asceticism has even penetrated ostensibly secular enterprises like science. Scientists and scholars are far from being free spirits. Actually, they are metaphysicians, for they still have faith in truth. Science rests on a metaphysical value, the absolute value of truth. Here as elsewhere in his remarks on science, Nietzsche is among the first to maintain that there is no such thing as a presuppositionless science. Science is only a *means*, and therefore presupposes certain values which give it direction. It is a metaphysical faith that underlies the value we place on science, a faith that goes back millennia to the Christian doctrine that God is truth, and beyond that to the Greeks who believed that truth is divine. Philosophers, both ancient and modern, have been oblivious of the fact that science itself requires justification. Philosophy's failure in this regard came about because it was dominated by the ascetic ideal, because truth was posited as the highest form of being (as in Plato) or as God (as in Christianity). In a word, the failure came about because truth was not allowed to become a problem (III, 24). The will to truth requires a critique, and the value of truth must be experimentally called into question; for once the God of the ascetic ideal is denied, a new problem inevitably arises, the value of truth. Thus Nietzsche advances the important insight that science itself never creates values, but, on the contrary, requires a value-creating power in the service of which it can be justified. It is not science, then, but *art* which, for Nietzsche, is fundamentally opposed to the ascetic ideal. For it is in art that the biologically-rooted, creative impulses express themselves most freely. Reflecting on the various epochs of history, Nietzsche speaks with disdain of those ages in which learned ascetic mandarins stepped into the foreground. In such ages the emotions grow cool, dialectics replaces instinct, seriousness is imprinted on faces and gestures, and life is impoverished. Nietzsche then goes on to make a statement that is fairly typical of his aristocratic utterances on politics: "The predominance of mandarins never signifies anything good; nor do the rise of democracy, *international peace courts in place of war*, equal rights for women, the religion of pity, and whatever other symptoms of declining life there are" (III, 25, italics added). The underscored passage in particular requires comment from those who have tried to save Nietzsche from himself by arguing that Nietzsche had intended his war metaphors and his praise of noble-warrior values to be taken in the sublimated, creative, artistic sense. Is there any good reason to deny that he had intended such passages to be taken literally? There is, however, good reason to suppose that as a nineteenth-century man he retained a view of war as a glorious event, an opportunity to re-invigorate the instincts and to breathe new life into the values of the "master morality."

Nietzsche closes the third essay of the *Genealogy* by reminding us of what he is trying to accomplish. From the time of Copernicus European man has been on an inclined plane, slipping faster and faster from the center; and for many Europeans the ascetic ideal has proved to be ineffectual in preventing them from falling into nothingness. The strongest spirits among them have discarded that ideal and have become atheists. Atheism, ironically, is the product of Christian morality's strict demand for the truth, but translated into scientific, rational terms. In a long and courageous act of self-overcoming, Christian dogma has been undermined by its own morality, by its own uncompromising demand for the truth. Christian truthfulness must end by drawing an inference against itself, which can only happen when it bravely faces the full implications of the will to truth (III, 27).

For Nietzsche, the curse that has lain over humanity is not suffering itself, but the fact that there has been no adequate answer to the crying question, "why do I suffer?" It is the meaninglessness of suffering, not suffering as such, that has caused humanity so much pain, and the ascetic ideal has offered humanity meaning, the only meaning offered so far. Suffering was interpreted in accordance with the ascetic ideal, and the tremendous void appeared to have been filled, forestalling suicidal nihilism. But the ascetic ideal was no unmixed blessing, since it brought with it a more inward, life-destructive suffering in the form of guilt, a bad conscience, and extreme forms of self-denial. Humanity has been temporarily saved by its willing of an ascetic ideal, Nietzsche acknowledges, but insofar as it is a rebellion against the most basic presuppositions of life, the ascetic ideal is a will to nothingness.

If we reflect on Nietzsche's writings considered so far, we can see that in his admiration for the master morality he leaves himself open to the charge that the will to power of the so-called higher specimens is beyond good and evil. That is tantamount to asserting that "might is right." How would one attempt to repudiate such an assertion? It may help us to think the problem through if we now consider how Socrates sought to refute it, and how he dealt with the proto-Nietzscheans of his time.

NOTES

1 John Dewey, *Human Nature and Conduct* (Southern Illinois University Press, Carbondale and Edwardsville, 1988), p. 84.
2 Charles Darwin, *The Descent of Man* (Princeton University Press, Princeton, 1981), pp. 156–7.
3 Sigmund Freud, *New Introductory Lectures on Psychoanalysis*, The Pelican Freud Library, tr. James Strachey (Penguin Books, New York, 1986), vol. 2, p. 107.
4 Sigmund Freud, *Standard Edition of the Comprehensive Psychological Works of Sigmund Freud*, tr. and ed. James Strachey (Hogarth Press and the Institute of Psychoanalysis, London, 1957), vol. 14, p. 282.
5 Ibid., p. 286.
6 Walter Kaufmann, *Nietzsche: Philosopher, Psychologist, Antichrist*, 4th edn (Princeton University Press, Princeton, 1974), p. 285.

7

Socrates and the Proto-Nietzscheans

From Plato we learn about a school of contemporaries who employed arguments from nature to oppose law, and who often proceeded to identify right with might. Plato records two forms of this doctrine, the first being more moderate and appearing at the beginning of the second book of the *Republic*. The second, more extreme form, appears in the dialogue *Gorgias* and in the first book of the *Republic*. The moderate form is stated by Glaucon, who is the elder brother of Plato, but not a Sophist:

> By nature, it is said, to commit injustice is good, and to suffer it is evil; but the evil is greater than the good. And so when individuals have both committed and suffered injustice, and have had the experience of both, being unable to avoid the one and take the other, they decide that they had better agree among themselves to have neither. Hence mutual covenants and laws arise; and that which is ordained by law is called by men lawful and just. (*Republic*, 358E–359A)

Glaucon then goes on to say that according to the theory in question, the best is to be able to do wrong with impunity and the worst is to be injured but lack the power to get one's revenge. Justice is midway between the two, and it is accepted and approved not as a real good, but as a thing honored in the absence of the ability to do injustice. For it is evident that he who possesses the power to do injustice – a real "man" – would never make a contract neither to wrong nor to be wronged (*Republic*, 359A–B). According to this view, there was an original condition of nature in which humans lived as individuals, each acting in accordance with his own will. Eventually, however, they entered into an agreement to give up the free exercise of their wills in return for the protection and preservation of their lives.

The second and more extreme form of the antithesis between nature and law is found in the *Gorgias*, where we witness the total rejection of the conventional justice instituted by a social contract, and a thoroughgoing advocacy of the natural right of might. This doctrine is ascribed by Plato not to Gorgias himself but rather to a certain Callicles, who rejects all law as the mere product of contracts made by the weak to defraud the strong of the just right of their power. Law, says Callicles – anticipating Nietzsche – institutes a slave-morality, which is no true morality, for nature and law are opposite, and nature is the true rule of human life. By nature, it is shameful to suffer wrong, but by convention it is more shameful to do it. Yet to suffer wrong is surely not fit for a man, but only for a slave, who is better off dead than alive, since when wronged he is unable to help himself or anyone else for whom he cares. For Callicles, those who frame the laws are the weak, the majority. They frame the laws for their own advantage, to prevent the stronger from dominating them. The weak try to frighten the strong by proclaiming that it is shameful and evil to overreach others, and by teaching that injustice consists in seeking an advantage over others. Although the weak are inferior, they nevertheless seek equality of status. Nature makes it plain, however, that it is right for the stronger and better to have the advantage over the weaker and the worse, the more able over the less. When Callicles argues from "nature," he has in mind both the animal world and the international arena – the latter being in a "state of nature," that is, a state of war in which the strong overpower and dominate the weak. Pointing to the behavior of animals and of states, Callicles insists that throughout "nature" it is the rule that the strong possess sovereignty and advantage over the weak. Callicles cites Xerxes' invasion of Greece as only one of countless instances in which men and states behave in accordance with the true nature of right, nature's own law. Callicles bemoans the fact that contrary to nature's own law, society molds and weakens the best and the strongest. Catching them young like lion cubs, society inculcates doctrines which make slaves of them, by teaching that they must be content with equality and that this is only right and fair. But, says

Callicles, if a man should arise with a nature sufficiently strong, he will burst his fetters, fling aside the domination of the herd, and break loose. He will trample on society's unnatural conventions as on so many insignificant scraps of paper, rise up, and reveal himself as our master instead of our slave (*Gorgias*, 483A–484C).

Thus inequality, for Callicles, is the rule of nature. It is only by convention that men can claim an equality of status or of distribution (*isonomia*), for by nature men are unequal, and the strong always gets more than the weak. The strength of which Callicles here speaks is, of course, not mere physical strength; it is the strength of both body and mind, including the force of *will* (*andreia*) backed by intellect (*phronesis*) (*Gorgias*, 491B, D). So we see definite affinities between the old Greek doctrine of the "will to power", including its anticipation of a "superman," and the teachings of Nietzsche. Like Nietzsche, Callicles is a moral revolutionary who throws aside the conventional or slave morality to make way for a master morality derived from the example of nature. There is such a thing as natural right, and its basis is might.

A still more extreme position than that of Callicles is represented by Plato as having been held by Thrasymachus of Chalcedon, a Sophist of the later fifth century. For Thrasymachus, there is no such thing as natural right. Right is simply whatever the strongest power in the state succeeds in enforcing in accordance with its own interests as it sees them. It does not matter what the strongest power in the state enforces, whether the advantage and interests of the strong or those of the weak, whether inequality or equality. Whatever it enforces is right. Thrasymachus does not subscribe to Callicles' doctrine that by nature's rule might is fundamentally right. Instead, Thrasymachus holds that right is nothing more than the establishment of power: if the weak (i.e. the majority or the people) make laws in their interest, those laws are just and right so long as the weak can enforce them; and the laws cease to be right the moment they cannot be enforced. While Callicles believes in a *natural* right which is always right, Thrasymachus believes there is no such thing as a single and permanent right. Thrasymachus believes that the only right is the enactment of the sovereign power. This is a form of ethical nihilism.

Let us, then, center our attention on Callicles' argument and observe how Socrates puts together his rebuttal. He begins by recapitulating the argument to make sure he understands it: "natural justice" according to Callicles and Pindar – whom he had cited in support of his view – means that the more powerful carry off by force the property of the weaker, the superior rule the inferior, the nobler take more than the meaner. Socrates then asks Callicles whether in his view the more powerful and stronger are also the better, and when he replies in the affirmative, that enables Socrates to lay a trap for him:

Socrates: Are not the many more powerful by nature than the one? And it is the many, as you yourself just stated, who frame their laws to control the one.

Callicles: Of course.

Socrates: Then the laws of the many are those of the more powerful?

Callicles: Yes.

Socrates: And of the better too? For you have asserted that the more powerful are also the better.

Callicles: Yes.

Socrates: Then their laws are naturally noble, since they are those of the most powerful.

Callicles: That's logical.

Socrates: Now, the many hold the opinion, as you yourself just stated, that justice means equal shares, and that it is more shameful to do wrong than to suffer wrong. Is that true or not? And take care that you yourself are not caught this time a victim of modesty. Is it the view of the many or not, that justice means equal shares, not unequal, and that it is more shameful to do than to suffer wrong? . . .

Callicles: Yes, that is the view of the majority.

Socrates: Ah, then it is not only by convention, but also by nature that it is more shameful to do than to suffer wrong; and it is also natural justice to share equally. So it appears that your previous assertion is not true and that you were mistaken when you insisted that convention and nature are opposed. . . .

Callicles: Will this fellow never stop babbling? Tell me, Socrates, are you not ashamed to play so cleverly with words at your age, taking advantage, of every misstep in your interlocutor's statements? Do you seriously imagine that by the more powerful I could mean anything but the better? Did I not tell you at the outset that the more powerful and the better are one and the

> same thing? Do you seriously think I could possibly mean that
> if a rabble of slaves and riff-raff – who are of no earthly use
> except for their physical strength – come together and make
> some proclamation, that is law?

To escape Socrates' trap Callicles shifts ground, and for the right of strength, which in the previous argument signified the right of quantity, he substitutes the right of *quality*. He now adopts the revised formula that those who are of better quality, or, in other words, of greater wisdom (*phronimoteroi*), should bear authority or rule. The Platonic Socrates has no objection to this formula so long as it is understood in a Socratic and not in an aristocratic sense, that "better" means morally better, and "wiser" means in philosophic knowledge. The formula is acceptable, moreover, only if it is taken to mean that the wiser has the right and duty to rule, but not the right to profit by his rule. Socrates makes this point in a parable. Suppose that a heap of food had to be distributed to a group of individuals, some of whom are strong and some weak. Surely we would want to assign the task of rationing to the most capable individual in their midst, namely, the physician, who is wisest where human bodily needs are concerned. But it would not follow that because he has the authority to distribute the food, he also has the right to receive a larger portion of it for himself.

Callicles, however, rejects this line of reasoning. When he said "wiser," he clarifies, he meant not only wiser, but also more manly and possessed of more strength of character (*andreioteroi*); and when he spoke of authority, he not only meant that intellectual power backed by strength of character should rule, but also that the bearer of authority should profit from ruling. The parable of the food is inept, for no man will take upon himself the burden of State affairs unless it is worth his while, unless he can profit from it personally. When Socrates replies that this view is hedonism, pure and simple, Callicles acknowledges that it is indeed hedonism, but goes on to say that

> the truth, Socrates, which you profess to pursue, is this: Luxury, in-
> temperance and license, when they have sufficient backing, are the essence
> of virtue and happiness; all the rest is trash, the unnatural catchwords of
> mankind, plain nonsense and of no account. (492C)

To this Socrates replies, in effect, that Callicles is blind to the fact that a life of self-gratification is a life of constant want. The selfish egoist tries futilely to fill a sieve; or in another metaphor, his life is like a torrent whose waters are always coming – and always going; or, finally, he is like a man who suffers from the itch and longs to scratch himself and who must scratch ceaselessly all his life. Can such a man be said to live happily (494B–D)?

Socrates now gives Callicles – an aspiring statesman just entering political affairs (500C and 515A) – a lesson in politics, his target being Athenian democracy. The orator-statesman of the kind Callicles emulates seeks not only to gratify himself, but also to please the multitude. He does so by flattering the people, thus forgetting the ethical injunction of his vocation, which is to leave his fellow citizens better men than he found them. Such a statesman fails to recognize his responsibility to instill in the hearts of the citizens the high virtues of balance, temperance, and justice. The true statesman is an artist totally committed to perfecting the object of his art; and for that he must be prepared to swim against the popular tide (505B–C). The old statesmen may have equipped the city with ships, fortifications, and arsenals (517C), but they failed in their primary responsibility of cultivating virtue. None of the founders of Athenian democracy escape Plato's condemnation; the original corruption goes back beyond Cimon to Themistocles and to Miltiades (503B–C; 516D–E). Even Pericles, the greatest figure of Athenian democracy, is charged with having satisfied the people's demands in order to satisfy his own ambitions (515D–516C). He gave the people pay for their public service and thus made them idle and greedy busybodies. That is the past of Athens; but today, says Socrates, every would-be statesman must ask himself whether he will be the physician of the state, striving to make its members as good as he can, or will be content to play the part of a servant and flatterer (521A). Thus Socrates who, like Nietzsche, detests democracy, nevertheless has no use for the moral relativism of the proto-Nietzscheans of his time.

One cannot do full justice to Socrates' rebuttal by confining oneself to the *Gorgias*, since Plato takes the whole of the *Republic* to develop the Socratic-Platonic conception of the relation of power to justice. It is there that Plato comes to grips with the ethical nihilism of Thrasymachus. Both Callicles and Thrasymachus are representatives of the revolt against traditional morality; both proclaim a new individualism which finds in the conventional morality nothing more than a number of limitations on bold, strong individuals. In the mouth of Callicles the new individualism is made to enunciate a new doctrine of "natural" justice – to seek whatever one likes and do whatever one can. The will to power of the stronger, "higher" individuals would not be obstructed if, instead of the conventions of the rabble, nature's rules were followed. The anticipation of Nietzsche's doctrine is obvious. In the mouth of Thrasymachus the new individualism becomes more cunning and gloomy: justice consists in obeying authority whenever one must, and in pleasing oneself whenever one can. And although the affinities are greater between Callicles and Nietzsche, we need to consider Plato's response to Thrasymachus in order to see more clearly how and why Plato's Socrates rejected the proto-Nietzschean position.

In the *Republic* Socrates seeks to expose the errors of extreme individualism by arguing that the self is no isolated unit, but an integral part of a social order. Thrasymachus had advanced two propositions: a government governs for its own advantage, and injustice is better than justice. In opposition to the former view Plato sets forth the Socratic conception of government as an art. All arts are called into being by defects in the material with which they deal. The physician attempts to remedy the defects of the body, the teacher those of the mind. The ruler or statesman, it follows, if he acts in accordance with his art, is absolutely *un*selfish; his single aim and commitment is to enhance the well-being of the citizens given over to his care. In opposition to Thrasymachus' second proposition Socrates replies with an argument designed to prove that the just man is a wiser, stronger and happier man than the unjust. He is wiser because he follows the old Delphic teaching and recognizes the need to acknowledge a limit. His aim is excellence; and if he asserts his individuality at all, it is in his striving for excellence. He does indeed compete with others but only incidentally, as it were, as he competes with himself in the pursuit of excellence. The just man is also stronger than the unjust man because he recognizes the principle that inevitably binds him to his fellows.

Everything, Plato maintains, has its appointed function (*ergon*) which cannot be fulfilled equally well by any other thing (352E). The specificity of things and their functions implies a division of labor in which everyone depends on everyone else. Hence, individuals engaged in discharging their respective functions do not compete or try to overreach one another because their respective functions are *complementary*, not competitive. The soul, too, has a definite function, with a corresponding virtue or excellence. Its function is life (*to zen*) and the corresponding virtue is good life (*to eu zen*). If the soul possesses the virtue of good living, or justice, it also possesses happiness (*eudaimonia*) which follows inevitably with good living. The just man, then, who strives for excellence in the pursuit of his appointed task in the division of labor, is happier than the unjust man.

These arguments are strictly logical in character and show us Plato playing with the Sophists at their game of words and trying to beat them at their own game. And the truth is that Plato's logic leaves us not wholly convinced. We are left with a nagging uneasiness that the Sophists have not been thoroughly trounced. We are also left with the impression that Plato's conception of justice is something to which human nature does not instinctively take, something unnatural, instilled in individuals by convention and held there by force.[1] Plato himself recognizes the inadequacy of his rebuttal and therefore amplifies his argument in response to the theory of Glaucon, who contends, in the spirit of Thrasymachus, that justice is an artificial thing, the product of convention. In the state of nature, says Glaucon, men injured and suffered injury without restraint. The weaker, seeing that they suffer more injury than they can inflict, make a "contract" with one another neither to do injustice nor to suffer it. They lay down a law – or "conventions" – which becomes the code of justice. As a result, human nature gives up its real instinctual urges, which demand self-satisfaction, and consents to be dominated by the force of law. For Glaucon, then, justice is the child of fear: "it is a compromise between the best of all, which is to commit injustice with impunity, and the worst of all, which is to suffer injustice without the power of retaliation" (*Republic*, 359A). Thus while Thrasymachus had founded justice on the instinct for domination and defined it as the *interest of the stronger*, Glaucon based justice on the instinct of fear and defined it as a *characteristic expedient of the weaker*.

Now, our aim here is not to decide whether Plato's Socrates or the sophistic "proto-Nietzscheans" got the upper hand. Our aim is rather to show that Socrates' rebuttal contains elements of what Nietzsche called a "slave morality," in that it attempts to meet the arguments of Callicles et al. by rejecting the view that man is by nature a selfish unit, and by proposing an opposite theory of human nature. Moreover, elements of the "slave morality" are evident in Socrates' inversion of Callicles' noble-warrior motto, that "might is right." Callicles asserted that by nature's rule one should practice overreaching others. He also reproached Socrates for holding a shameful view that leaves him incapable of helping himself, his friends, and relatives, and at the mercy of anyone who wishes to box his ears, steal his money, drive him out of the city, or even put him to death. Socrates, however, responds to Callicles as follows:

> I hold the view, Callicles, that it is *not* the most shameful of experiences to be wrongfully boxed on the ears, nor to have either my purse or my person cut, but it is more disgraceful and, indeed, wicked, wrongfully to strike or to cut me; and, furthermore, that theft, kidnapping and burglary and, in a word, any wrong done to me and mine is at once more shameful and worse for the wrongdoer than for me the sufferer. (*Gorgias*, 508D–509E)

Indeed, Socrates goes even further in inverting the "master morality." In the dialogue *Crito*, where a friend by that name tries to persuade Socrates to escape from prison in order to avoid execution, Socrates maintains that there is no instance in which injuring others is good or honorable. In no circumstances must one do wrong, not even when one is wronged – which most people regard as the natural course. There is no difference between injuring people and wronging them; "it is never right to do a wrong or return a wrong or defend oneself against injury by retaliation . . ." (*Crito*, 49A–E). Here we see Socrates anticipating "turn the other cheek," and espousing a view that unequivocally repudiates the "master morality." That is one reason for Nietzsche's rather negative attitude towards Socrates' teachings.

There is, however, another reason as well. Socrates believed in the eternity of the soul; and his abiding concern was how his soul will fare in the other world. In his debate with Callicles, Socrates *reasons* with him, of course, and seeks to persuade him that his argument from nature is wrong. But when Callicles persists, saying that it is disgraceful for a man to be unable to defend himself against injury and death, and reproaching Socrates for espousing such an unmanly philosophy, Socrates replies that

> no one who is not utterly irrational and cowardly is afraid of the mere act of dying; it is evil-doing that he fears. For to arrive in the other world with a soul heavy with many wicked deeds is the worst of all evils. And if you like, I can tell you a tale which will prove that that is true. (*Gorgias*, 522D–E)

In that tale Socrates relates that in the days of Cronos there was a law concerning humanity which from that time to the present has prevailed among the gods: the individual who has led a righteous life departs after death to the Isles of the Blessed and there lives in happiness safe from all ills, but the unrighteous individual departs to the prison of punishment which is called Tartarus. In the last days of Cronos, however, when Zeus had just come to power, Pluto and the stewards of the Isles of the Blessed informed Zeus that it was living men who decided the verdicts on other living men on the very day that the latter were to die, and that the wrong people were being sent to both places. So Zeus changed this unjust state of affairs and appointed his sons as judges.

Callicles is then informed by Socrates that this is the story he has heard, that he believes it to be true, and that he draws from it the following conclusion: death is nothing other than the separation of the soul from the body, and just as a corpse retains all the blemishes and scars and other marks attesting to the experiences of that body when it was alive, the same is true of the soul. Once it has been separated from the body, all of the experiences which an individual's soul has encountered are manifest, as are its characteristics. When, therefore, the souls arrive before their judge, he scrutinizes them with care – but with even greater care does he scan the souls of former kings and potentates. And when the judge discerns that due to crimes and perjuries the soul is full of scars – the marks branded upon it by every evil deed it has perpetrated – he sends it in ignominy straight to the prison-house in Hades, where it is destined to endure the sufferings it has earned. Those who are improved through punishment prove to be curable; but those who have committed the most heinous crimes, and whose misdeeds are past cure, serve as warnings to the others, who witness the suffering of the incurable throughout eternity. The incurable endure the most excruciating and terrifying tortures, literally suspended there in the prison-house, a warning to any evil-doers who from time to time arrive. And Socrates adds:

> I think that most of those warning examples are selected from tyrants and kings and potentates; for, owing to the license they enjoy, they are guilty of the worst and most impious crimes.

And he underscores for Callicles that

> it is among the most powerful that you find the superlatively wicked. (*Gorgias*, 523A–525E)

Callicles is assured by Socrates that he is convinced by these stories and that he always considers how he may present to his judge the healthiest possible soul by pursuing the truth and endeavoring to live out his life as good a man as he can possibly be. And he exhorts others to do the same. Only those who lack wisdom would live any other life than this, since it is clearly of benefit not only in this world, but in the other world as well. In closing, he reminds Callicles again that one should be more on guard against doing than suffering wrong; and he assures him that even if someone strikes him a so-called humiliating blow, he should accept it cheerfully, for he will suffer no real harm thereby if he really is a good man and pursues virtue (*Gorgias*, 526E–527D).

From this brief survey of Socrates' encounter with the proto-Nietzscheans, one can see why Nietzsche would have had an unfavorable attitude toward Socrates. For there can be no doubt that his philosophy partakes of the slave morality and repudiates the master morality based on arguments from nature. In addition, Socrates is not only the author of the metaphysical theory of the Forms, he is a religious man who believes in the eternity of the soul and in a judgment-day in the "other world" where the soul will be directed either to the Isles of the Blessed or to the prison-house in Hades. So Nietzsche, the thoroughgoing materialist, cannot admire Socrates either for his metaphysics or for his participation in the Greek inversion of the noble-warrior values. Moreover, as we shall see in *Twilight of the Idols*, Nietzsche is positively hostile towards Socratic dialectics.

Walter Kaufmann has, however, insisted that Nietzsche's attitude toward Socrates was essentially positive. Kaufmann attempts to defend this view by relying largely on Nietzsche's first book, *The Birth of Tragedy*. There, we recall, Nietzsche posited destructive consequences for art flowing from the influence of Socratic rationality: when the older duality, Dionysian-Apollinian, was superseded by the new opposition, Dionysian-Socratic, "the art of Greek tragedy was wrecked on that account" (section #12). The supreme law of *aesthetic Socratism*, says Nietzsche, reads as follows:

"Everything must be understandable in order to be beautiful." This is the corollary of the Socratic principle that "knowledge is virtue." With this canon in his hands, Euripides measured all the separate elements of drama – language, characters, dramatic structure, choral music – and corrected them according to that principle.

In comparison with Sophocles, the poetic deficiency and degeneration [*Rückschritt*] which are so often ascribed to Euripides are for the most part the product of that penetrating critical process, that audacious rationality. (section 12)

Moreover, with regard to our thesis that Socrates was an active participant in the inversion of noble-warrior values, it is noteworthy that he was perceived as such by his contemporaries, which is one reason why he was lampooned by Aristophanes in his comedy, *The Clouds*. The Athenians, Nietzsche observed,

attributed to the influence of Socrates and Euripides the fact that the old Marathonian hardy fitness of body and soul was increasingly being sacrificed to a dubious enlightenment which entailed the progressive degeneration of the powers of body and soul. It is in this tone, half indignant, half scornful, that Aristophanic comedy used to speak of both of them – to the dismay of modern men who are quite willing to give up Euripides, but who cannot adequately express their astonishment that Socrates should appear in Aristophanes as the first and supreme *Sophist*, the mirror and embodiment of all sophistical tendencies. (section 13)

Writing from an aesthetic point of view, Nietzsche criticizes Socrates for failing to recognize the creative force of instinct in art, and for replacing instinct with reason as the creative principle. Socrates was blind to the frenzy of artistic enthusiasm so necessary to the creative artist.

In his first book, then, it is true that Nietzsche speaks of the "divine naïveté and sureness of the Socratic way of life", and that he also speaks with admiration of the way Socrates brought the death-sentence upon himself with full awareness and without the natural fear of death (section 13). On the other hand, it is also true that the young Nietzsche – 28 years of age when he published *The Birth of Tragedy* – is already quite skeptical of the powers of reason as employed by Socrates and Plato. Already here, Nietzsche, the aphorist of aestheticism, has embraced Dionysus at Socrates' expense; for Socrates was the first to foster a profound illusion,

> the unshakable faith that thought, employing the guiding thread of causality, can reach down into the deepest abysses of being, and that thought is not only capable of knowing being but *of correcting* it as well. This lofty metaphysical illusion instinctively accompanies science again and again to its limits, at which it must turn into *art* . . . (section 15)

When, therefore, Walter Kaufmann characterizes Nietzsche's attitude toward Socrates as one in which he pays only homage to the first "theoretical man," this characterization tends to gloss over the view ostensibly held by Nietzsche, that the power of reason is a profound illusion. In his first book as in his entire *oeuvre*, his homage to Socrates pales in comparison with his homage to the Dionysian will to power.

NOTE

1   A fuller critique of Plato's theories may be found in Irving M. Zeitlin, *Plato's Vision: The Classical Origins of Social and Political Thought* (Prentice Hall, Englewood Cliffs, N. J., 1993).

8

An Excursus on Max Stirner – and Karl Marx

In the modern era the pre-Nietzschean thinker who most conspicuously followed in the footsteps of Callicles and Thrasymachus was a man who was christened Johann Caspar Schmidt, but whose unusually high forehead gained him at school the nickname of "Stirner." In time this so tickled his fancy and stirred his ambitions (*Stirn* = forehead; *Gestirn* = star) that by the time he entered the university at Berlin he had abandoned the plebeian name of Schmidt and become "Max Stirner." At Berlin his philosophical consciousness was awakened by Hegel, and he soon met other admirers of the famous philosopher who called themselves Young Hegelians and who assembled frequently in Hippel's *Weinstube* (tavern) on Friedrichstrasse. This circle of radical neo-Hegelians included Karl Marx, Frederick Engels, Arnold Ruge, Bruno and Edgar Bauer, and others.

It is Stirner more than any other thinker of the early nineteenth century who deserves the epithet "proto-Nietzschean"; and it is highly probable that Nietzsche knew Stirner's work and that he was profoundly influenced by it. The degree to which Nietzsche is anticipated by Stirner both in ideas and in prose style can hardly be a coincidence. Their central concerns are too similar as are their key concepts: antichrist, immoralism, priest-morality, irrationalism, and egoist-superman. Stirner also wrote of the "death of God" and the decadent nature of democracy. The proto-Nietzschean book for which Stirner is best known is entitled *Der Einzige und sein Eigenthum*, which has been translated as

The Ego and His Own.[1] This book first appeared in 1844, within days of Nietzsche's birth. Stirner is nowhere mentioned in Nietzsche's published writings. However, C. A. Bernoulli, in a book on Nietzsche and his friend Franz Overbeck, relates the story of another of Nietzsche's friends, Adolf Baumgartner, to the effect that during a semester at Basel in 1874, which he had spent in Nietzsche's company, he read Stirner's book at Nietzsche's "warmest recommendation." And John Carroll writes in this regard:

> It has been confirmed that Baumgartner, but not Nietzsche, borrowed *The Ego and His Own* from the Basel library, on July 14th of the same year [1874]. There are a number of sources from which Nietzsche could have obtained the book, most likely in the late 1860s. Overbeck's final conclusion after finding more, quite persuasive, but circumstantial, evidence was that Nietzsche had read Stirner, was impressed, and worried that he should be confused with him. It is true that Nietzsche showed a virtually obsessive concern for originality.[2]

Stirner's egoistic motto is "Realize yourself!" which is closely akin to Nietzsche's "Become who you are!" – in sharp contrast to Socrates' "Become just!" Stirner likens himself to a singer in the heights of his song, who sings for his own sake, not for that of anyone else, and not for the sake of truth. Like Nietzsche after him, Stirner also believed that in music the passions of the soul were stirred, gaining supreme expression and gratification. Stirner also employs an aphoristic style and founds his project on the rejection of the traditional concerns which actually amount to *nothing* – the Good, the Divine, the so-called Cause of Mankind, the Truth, Freedom, Justice, the People, the Fatherland. He himself is his only concern, and in that concern there is neither good nor evil, for neither of these have any real meaning for him. As everyone cares only for himself, he comes into conflict with others and the battle of self-assertion is inevitable. Hence, victory or defeat, that is the question for everyone. The victor becomes the lord, the vanquished the subject, the victor exercising supremacy as a right, the vanquished the duties of a subject. They become and remain enemies, each looking out for the other's weaknesses. Either the rod conquers the man, or the man conquers the rod. But behind the rod, mightier than it, stands courage.

As children our instincts are vital, and the feeling of self strives to defer to nothing and no one. Soon, however, our impulses and our inclinations meet with the objections of the mind, the objections of one's own conscience. The notion emerges that certain actions and thoughts are bad, unreasonable, unchristian. Boys give free play to their impulses and have only unintellectual interests; youths often weaken their impulses with interests almost exclusively intellectual. The *real* man, however, has only bodily, personal, and egoistic interests. Only when a man has fallen in love with his corporeal self and has derived pleasure in himself as a living, flesh-and-blood individual, has he acquired a *selfish* interest so necessary for self-realization (*Ego* 3–16).

Stirner wants to reject all metaphysics, not only notions of God but notions of "Man" as well, which have been mistakenly rendered sacred by the materialists. But actually it is only the individual striving to raise himself above the limits of his individuality that is real. He exists only in raising himself and not remaining as he is – thus anticipating Nietzsche's "self-mastery" or "self-overcoming." Man with a capital M is an ideal, which is the same as saying it is nothing. For Stirner, there is no essential difference between looking to "humanity" as an ideal and looking to God and Christ.

Nietzsche's superman–ascetic priest dichotomy parallels Stirner's egoist–cleric contrast at many points, as does its central assumption that modern man's malaise stems from moral rather than from economic or political repression. In these terms Stirner has a stronger claim than Nietzsche to be regarded as the first immoralist in modern times. Each individual, says Stirner, carries stern moral injunctions in his breast; hence, he cannot *will* to gain freedom, he can only wish or petition for it. What would happen if the "slave" really willed with the full energy of his will? But no! – instead he renounces will for the love of morality, which is why he is condemned to remain a slave. Even where enlightenment and rationalism have challenged and weakened conventional morality, courage is lacking to devote oneself wholly to egoism. Throughout history opposition parties have failed for lack of will. Take Nero, in whom the "good" saw an arch-villain. Why did people put up with him for so long? One must not suppose that those "tame" Romans of the time, who allowed their will to be bound by such a tyrant, were one iota better than he. In old Rome they never would have remained his slaves for long. But those "good" contemporaries of the tyrant opposed him with moral demands, not their will. Self-owning men, however, would never shamefully and foolishly whine in the face of a tyrant's disregard for the "sacred" law; they would oppose him with their will and their power.

Stirner also anticipates Nietzsche in denouncing self-renunciation and the repression of the instincts. There sits youth with its deathly tired head and pale cheeks. Poor children, the passions beat at their hearts, and the rich powers of youth demand their right. Nature quivers in their limbs, the blood swelling in their veins, fiery fancies exciting the gleam in their eyes. But then the conscience appears, and the young innocents turn their terrified eyes upward and they pray. The storms of nature are hushed and the surging sea of their appetites is calmed, but only apparently so. The soul is saved, but, alas, the body may perish. Human beings have continually sought a standpoint *outside* of the earth and have found it in the world of the mind – in ideas, concepts, essences and "heaven." Heaven is the unreal standpoint from which the earthly doings of humans are looked down upon and despised. Christianity has thus succeeded in "delivering" us from a life determined by nature, from the appetites impelling us. Humans, according to this ascetic doctrine should not allow themselves to be actuated by their appetites. Stirner wants to apply to *mind* (faith, ascetic moral teachings) what Christianity has contrived against the instincts: if, as Christianity teaches, we are indeed to have appetites but the appetites are not to have us, then we should now say we are indeed to have mind, but the mind is not to have us. It is only through the "flesh" that one can break the tyranny of the mind; for it is only when an individual hears his flesh along with the rest of his physical self that he hears himself wholly.

The idea of the "sacred" renders people powerless and humble. And yet nothing is sacred in itself. It is one's declaring it sacred, one's bending of the knee that makes it so. "Sacred" is a phenomenon of consciousness and conscience; it is everything that stands before and "above" the egoist as an obstacle to his will to life. For little children as for animals, the sacred has no existence. Only later, as children learn the distinctions "good and bad," do reverence and "sacred dread" step into the place of natural fear. That is the standpoint of religion – holding something outside oneself as mightier, greater and higher. It is the attitude in which one projects onto an alien, hypothetical something such powers that one expressly yields and surrenders oneself to it in devotion, humility, servility, and submission. Religion is the phenomenon that bars the way to an individual's self-will. Instilling fear in an individual is the attempt to raise a barrier against his egoistic self-will. But fear is not enough, since there always remains the effort to overcome what is feared by means of guile and deception. When something is regarded as sacred, however, it is not only feared but revered and honored. With reverence one has internalized a power from which one no longer strives to liberate oneself; one is attached to it with all the strength of faith. One believes! Even for non-believers, however, morality itself has acquired a sacred status. Moral concepts in and of themselves are revered. From infancy one is indoctrinated with the notion that one must be moral. Whether morality itself is an illusion is a question no one dares to address; it remains exalted above all doubt. Spiritual men have taken it into their heads to realize the *concepts* of goodness, of love. They want somehow to establish the kingdom of love on earth, in which no one any longer acts from selfishness. This *idée fixe* will surely lead humanity to ruination.

Stirner now introduces the concept of "clericalism" to denote one who lives for a "lofty" idea, cause, doctrine or system which prevents all worldly lusts and self-seeking interests from springing up in him. Stirner appears to be among the first of modern thinkers to use "cleric" (*der Pfaff*) to symbolize the resentful, inhibited, devitalized perpetrator of morality, whose teachings paralyze the will and deprive it of any inner authority. "Cleric," in these terms, anticipates Nietzsche's "ascetic priest" (*asketischer Priester*). "Clericalism" implies that one feels oneself called to live and work for an idea, or a "good cause," and to renounce all pleasures and enjoyments that flow from sources other than the "sacred interests." The "sacred interests" are not only the traditional, religio-moral interests, but also, for example, those of Fatherland, Science, and Revolution. Stirner thus assails the secular ideals as well, arguing that he who is infatuated with *Man* ". . . leaves persons out of account so far as that infatuation extends, and floats in an ideal, sacred interest. *Man*, you see, is not a person, but an ideal, a spook" (*Ego* 97–101).

Like Callicles, and before Nietzsche, Stirner envisions a radical break with the past as a result of which strong-willed individuals will arise and no longer let anything be whined or chattered into them by the "clerics." The new and higher types will have no sympathy for the follies which the "clerics" have been raving and drivelling about since the beginning of history. The new strong-willed individuals will destroy original sin and all the other debilitating stupidities inherited from the past. "If you command them," writes Stirner,

> "Bend before the Most High," they will answer, "If he wants to bend us, let him come himself and do it; we, at least, will not bend of our own accord." And if you threaten them with his wrath and his punishment, they will take it like being threatened by the bogeyman. If you are no longer successful in making them afraid of ghosts, then the dominion of ghosts is at an end, and fairytales find no – faith. (*Ego* 105–6)[3]

The new strong-willed individuals will also see through the latest metamorphoses of the Christian religion such as humanism, liberalism, democracy, and communism. All of these human religions exalt "Man" to the same extent that any traditional religion exalts its idol or god, turning "Man" into something otherworldly and alien and setting real human beings beneath it. Nietzsche was not the first to cry out, "God is dead!" and "we have killed him" (*The Gay Science*, section 125). Much earlier Stirner had made the same proclamation, and went on to say that the job is only half done:

> At the entrance to the modern time stands the "God-Man." At its exit will only the God in the God-Man evaporate? And can the God-Man really die if only the God in him dies? They did not think of this question, and thought they were through when in our days they brought to a victorious end the work of the Enlightenment, the vanquishing of God: they did not notice that Man has killed God in order to become now – "*sole* God on high." The *other world outside us* is indeed brushed away, and the great undertaking of the Enlighteners completed; but the *other world in us* has become a new heaven and calls us forth to renewed heaven-storming: God has had to give place, yet not to us, but to – Man. How can you believe that the God-Man is dead before the Man in him, besides the God, is dead? (Part Two, I)[4]

Nietzsche's concern with the "bad conscience" and its repression of the human drives and instincts is also first raised as an issue by Stirner. The human individual is a bundle of impulses, desires, wishes, and passions. From the beginning the traditional view has asked: how are these to be correctly regulated if not in obedience to God's commandments? And if not in obedience to God, then at least in obedience to the moral principles and the voice of reason which in the course of the many bitter experiences of history have been raised into law? Wouldn't the passions lead us to do destructive and senseless things without the light of a guiding star? Stirner's reply is much the same as that of Callicles and Nietzsche. "If the individual only deemed himself a beast," writes Stirner,

> he would easily find that the beast, which does follow only *its* impulse (as it were, its advice), does not advise and impel itself to do the "most senseless" things, but takes very correct steps. But the habit of the religious way of thinking has biased our mind so grievously that we are – terrified at *ourselves* in our nakedness and naturalness; it has degraded us so that we deem ourselves depraved by nature, born devils. (*Ego* 211–13)[5]

Thousands of years of civilization have obscured from us what we actually are, deceiving us into thinking that we are good when we deny ourselves. Shake that off, says Stirner, and become the open egoists that your nature intends you to be. Become an *almighty ego*!

Like Callicles and Thrasymachus before him, Stirner proposes that "might is right." The ego secures its freedom insofar as it makes the world its own, takes possession of it for itself, "by whatever might" – persuasion, categorical demand, even by means of hypocrisy and cheating. The ego's freedom is diminished when it cannot realize its will on another object, whether it be an object without will, like a rock, or with will, like another individual or a state. One circumvents a rock obstructing one's way until one has powder enough to blast it; one submits to the laws and authorities of the state until one has gathered enough strength to overthrow it. "My freedom," writes Stirner,

> becomes complete only when it is my – *might*; but by this I cease to be a merely free man, and become an own man. Why is the freedom of the peoples a "hollow word"? Because the peoples have no might? With a breath of the living ego I blow peoples over, be it the breath of a Nero, a Chinese emperor, or a poor writer. Why is it that the German legislatures pine in vain for freedom, and are lectured for it by the cabinet ministers?

Because they are not of the "mighty"! Might is a fine thing, and useful for
many purposes; one goes farther with a handful of might than with a
bagful of right. You long for freedom? You fools! If you took might,
freedom would come of itself. See, he who has might "stands above the
law." How does this prospect taste to you, you "law-abiding" people? But
you have no taste. (*Ego* 214–23)[6]

Freedom that is granted or given is no freedom at all. Only the freedom
one *takes* for oneself is the real egoist's freedom, the product of will so
gratifying to one's passions. When the Greeks came of age they drove out
their tyrants and did not wait for the tyrants to grant them their freedom.

My ego is the only judge of whether I am in the right or not. Others
can only decide whether to endorse my right. To the question, what or
who gives me the right to do this or that, the traditional answer is God,
love, reason, humanity, and so on. But in reality it is only one's own
might that gives one a right. There is no right outside of me. If it is right
for me, it is right; and if that does not suffice to make it right for others,
that is their problem, not mine. Let them fight and defend themselves
against my right. If something is right for me and I want it, though it is
not right for the whole world, then I neither ask the world for its opinion
nor care about it. That is the rule for every strong individual who knows
how to value himself, "for might goes before right, and that – with
perfect right" (*Ego* 247–251).[7] Formerly, divine reason was placed above
and against human reason; now human reason, so-called, is placed above
and against egoistic reason, which is rejected as "unreason." Little do
people realize that neither divine nor human reason is real, and that the
only reason is that of the individual. The strong-willed individual is the
irreconcilable enemy of every generality, every tie, every fetter.

Property, for Stirner, is essential for the self-realization of the individual. To what property is the individual entitled? Only to that which
is in his power to obtain. One gives oneself the right to property by taking
it to oneself. Might, again, is the key: everything over which one has
might, and which cannot be torn from him, remains his property.
The individual claims as property everything he feels himself strong
enough to obtain, and he extends his actual property by as much as he is
empowered to take. Egoism and selfishness decide the issue, not justice,
not mercy, not goodness, not love, nor any other form of selflessness.

Egoism never thinks of sacrificing or giving away anything it wants. It simply decides that what it wants it must have and will acquire. "Take hold," Stirner proclaims,

> and take what you require! With this the war of all against all is declared. I alone decide what I will have. (*Ego* 333–42)[8]

Stirner thus repeats the phrase Hobbes employed to describe the "state of nature." Whereas Hobbes, however, advanced the valid proposition that a strong central power is necessary in order to bring to an end the "war of each against all," Stirner turns the exercise of power and coercion into a principle for individual action. From a theoretical standpoint, then, it appears that for Stirner human *individuals* always remain in a "state of nature," as do States in the international arena. Is this equally true of Nietzsche's theory, insofar as it is a *political* theory and not merely an aesthetic doctrine, as some might argue? As we observed earlier, Nietzsche never faces up to the Hobbesian problem: if the "will to power" is to become the principle of individual action, how is a "war of each against all" to be prevented? Or would he agree with Stirner that we have been and continue to remain in a "state of nature," and that it is only with the coming of the "higher specimens" that the war will come to an end with the establishment of a truly worthy and mighty Leviathan consisting of supermen?

Stirner goes quite far in rejecting "community" and advocating one-sided egoism. He urges his readers not to dream of the most comprehensive commune, "human society," but rather to see in others only means to one's own egoistic ends. No one, says Stirner, is his equal; for everyone, without exception, is his property. Much of what Stirner has to say here is directed against Bruno Bauer and some of the other young Hegelians in his circle who had proposed that one should be a man among "fellow men," that one should respect the fellow-man in all human beings. To this Stirner replies:

> For me no one is a person to be respected, not even the fellow man, but solely, like other beings, an *object* in which I take an interest or else do not, an interesting or uninteresting object, a usable or unusable person. (*Ego* 414–17)[9]

Stirner's philosophical aim is to teach that in one's relation with the world, one wants the enjoyment of it. Therefore the world must become one's property and one must want to gain it. Stirner does not shy away from taking this doctrine to its logical conclusion:

> ... *my* satisfaction decides about my relation to man, and ... I do not renounce, from any excess of humility, even the power of life and death.[10]

Stirner's open advocacy of man's making himself God and taking upon himself the power to murder is an idea that both fascinated and terrified Dostoevsky. *Crime and Punishment* represents the great writer's impassioned pondering of the implications of a man, smitten by Enlightenment ideas, who sets out to prove himself a "higher being" through the premeditated murder of a miserly old woman whom he *reasons* to be worthless. Dostoevsky's brilliant treatment of the phenomenon and Stirner's profound influence on the Russian intelligentsia, will be fully discussed in chapter 12.

In one's relation to the world Stirner holds that one should no longer do anything "for God's sake," or for humanity's sake, but only for one's own sake. Stirner's philosophy represents the most provocative pre-Nietzschean assault upon Western values in the modern era. It is an endorsement of egoism, hedonism, and the doctrine "might is right," taken to its logical extreme. It may therefore be of interest to listen to two of Stirner's contemporaries who also regarded themselves as "materialists," but who rejected Stirner's doctrine with an argument similar to that employed by Socrates in his response to Callicles and Thrasymachus. We have chosen the rebuttal of Marx and Engels because, like Stirner, they were members of the Young Hegelian circle and because they devoted more than half of their book, *The German Ideology*, to a systematic critique of Stirner's philosophy.

THE REBUTTAL OF MARX AND ENGELS

With the exception of Stirner, all of the Young Hegelians believed that classical German philosophy had made an epoch-making contribution to knowledge by underscoring the *social* nature of consciousness. What

Kant, Fichte and Hegel had expressed in the language of philosophical idealism, the Young Hegelians formulated in more historical, sociological and psychological terms: the "individual" is an abstraction; there is no self without an other; individual rights and obligations are social and historical, and must be analyzed and understood as such. Stirner, however, who could find in his head only his own self and consciousness and no other ego, regarded the emphasis on the social nature of the mind as a serious threat to individual freedom and the autonomy of the personality.

Marx and Engels begin by acknowledging some of the positive merits of Stirner's work, such as the honesty in his contention that one should consider oneself. This contrasted with the less-than-honest, sickly altruism of several contemporary writers who exhorted their readers to consider only others. Positive, too, was Stirner's insistence that freedom must mean freedom to act: where one is nominally free but lacks the material means with which to realize oneself, one is actually unfree. As a materialist, Stirner claimed that he had rejected idealist philosophy in its entirety. To this Marx and Engels replied that Stirner had merely rejected specific ideals such as God, immortality, and humanity. But he had unsuspectingly retained the idealist *method*, which disregarded the historical and social context of ideals and which sought to discover one absolute ideal through abstract logical or psychological analysis. Hence Stirner had inadvertently replaced the abstractions of religion and speculative philosophy – God and Man – with an even more preposterous abstraction, the "Ego." What is the "self?" Marx and Engels ask. Is it not an abstraction from a whole complex of social relationships, of selves in relation to one another? In the words of Marx and Engels:

> Individuals have always and in all circumstances "started out *from themselves*," but since they were not unique in the sense of not needing any connections with one another, and since their *needs*, consequently their nature and the method of satisfying their needs, connected them with one another (relations between the sexes, exchange, division of labor), they *had* to enter into association with one another not as pure egos, but as individuals at a definite stage of development of their productive forces and requirements, and since this association, in its turn, determined

production and needs, it was, therefore, precisely the personal, individual behavior of individuals, their behavior to one another as individuals, that created the existing conditions and daily reproduces them anew.... Hence it certainly follows that the development of an individual is determined by the development of all the others with whom he is directly or indirectly associated, and that the different generations of individuals entering into relation with one another are connected with one another, that the physical existence of the later generations is determined by that of their predecessors, and that these later generations inherit the productive forces and forms of association accumulated by their predecessors, their own mutual relations being determined thereby. In short, it is clear that a development occurs and that the history of a single individual cannot possibly be separated from the history of preceding or contemporary individuals, but is determined by this history.[11]

Marx and Engels then go on to argue that in the era in which they and Stirner live, the prevailing structure of economic relations is such that the

suppression of individuality by chance has assumed its sharpest and most universal form, thereby setting existing individuals a very definite task. It has set them the task of replacing the domination of circumstances and of chance over individuals by the domination of individuals over chance and circumstances. It has not, as Sancho [Stirner] imagines, put forward the demand that "I should develop myself," which up to now every individual has done without Sancho's good advice; it has instead called for liberation from one quite definite mode of development.[12]

The heart of this rebuttal, then, is the recognition of the fundamental interdependence of the self and the other, and the impossibility of separating them outside of one's imagination. Stirner's notion of "one's own" is therefore a highly artificial abstraction from the interdependent nature of all individual activities in any society. Every individual, without exception, is either directly or indirectly dependent on others for his or her livelihood; and every individual is dependent upon the accumulated store of social knowledge and material. No individual can therefore justifiably make the claim that "his own" was acquired solely through his own efforts. This has definite implications for Nietzsche's

conception of the "higher type." For even when a higher, creative individual works alone, it is undeniable that much more than his own
"unaided" efforts have gone into the production of the specific goods he
has created – the much more being the techniques, traditions, language,
and knowledge without which he could do nothing. This principle
applies inescapably to those whom Nietzsche envisions as the "supermen" or "higher types." For no matter how gifted, creative, powerful, or
"superior" an individual might be, he owes his achievements, in a large
measure, to others. In a letter Engels wrote to Marx on November 19,
1844, he criticized Stirner but acknowledged that

> what is true in his principle, we, too, must accept. And what is true is that
> before we can be active in any cause we must make it our own, egoistic
> cause – and that in this sense, quite apart from any material considera-
> tions, we are communists in virtue of our egoism, that out of egoism we
> want to be *human beings* and not merely individuals. [13]

And, of course, Engels and Marx did in fact devote their "own" to a
cause in which they envisioned "the free development of each leading to
the free development of all." For our purposes, the point of all this is that
although Marx and Engels, as philosophical materialists, also believed
that "God is dead," they nevertheless adopted and endorsed the values of
the "slave morality" and gave their own to the oppressed of their time.
In Engels' words, they wanted, out of egoism, to be human beings and
not merely individuals. Nietzsche, however, appears to overlook the
fundamental principles of "interdependence." Like Callicles and Stirner
before him, he draws on analogies from nature to posit a radical separation and antagonism between the "higher type" and the "herd." And
since he holds this view largely on the basis of his reading of Darwin, it
will repay us to have another close look at Darwin's theory.

NOTES

1   Ed. and introduced by John Carroll (Jonathan Cape, London, 1971). The English
    translation by Steven T. Byington, from which the extracts quoted in this chapter
    are taken, was first published in 1907. The present edition by John Carroll contains
    large excerpts of Stirner's massive work.

2   Ibid., p. 25.
3   Ibid., p. 84–5.
4   Ibid., p. 109.
5   Ibid., p. 116.
6   Ibid., pp. 121–2.
7   Ibid., p. 127.
8   Ibid., pp. 167–8.
9   Ibid., p. 214.
10 Ibid., p. 222.
11 Karl Marx and Frederick Engels, *The German Ideology* (Progress Publishers, Moscow, 1964), pp. 481–2.
12 Ibid., p. 482.
13 Cited in Sidney Hook, *From Hegel to Marx* (The University of Michigan Press, Ann Arbor, 1962), pp. 173–4 (italics added).

9

Darwin contra *Nietzsche*

Darwin himself had never portrayed the animal world as engaged in a ferocious, dog-eat-dog struggle for existence. Nor did he depict that world as one in which the stronger or better adapted individuals either dominated the herd or separated from it. Nor, finally, did he ever suggest that, because they have evolved from "lower" animals, human beings have created a world of hostility, conflict, and aggression. As we observed earlier, Nietzsche accepted the validity of Darwin's theory and understood it well in most respects. He does appear, however, to have missed the significance of Darwin's work for his own philosophy.

Darwin's thesis is based on the fact that offspring differ physically from their parents in significant ways. Those changes, which he termed *variations*, are today called mutations. Darwin assumed that the new traits best suited to the circumstances in which a species lives will be most likely to appear again in the next generation. In that way the most successful variations will be transmitted from generation to generation until the species gradually evolves, through the process of natural selection, into a somewhat different species. That is also the way humans evolved out of earlier animal forms. To illustrate the process of natural selection with just one example, let us assume that the existence of a thinly-furred species is gravely threatened as the climate grows colder owing to the advent of a glacial period. As a result of a mutation, some members of the species acquire a thicker growth of fur. They are more likely to survive and reproduce than the less hairy animals. Some of their offspring, in turn, would adapt more easily to the increasing cold and would transmit the thick-furred genes to the next generation. Thus gradually, over a period of many generations, an entire species might change from being thinly to thickly furred.

Clearly, the process of natural selection entails no fierce struggle for survival *among members of the same species*. It is a "struggle for existence," as Darwin termed it, only in the sense that every organism strives to maintain itself. Darwin himself emphasized in *The Origin of Species* that he employed this phrase in a large and metaphorical sense, including the

> *dependence of one being on another*, and including . . . not only the life of the individual but success in leaving progeny.[1]

And in *The Descent of Man* Darwin went on to show that in numerous animal species the struggle between individuals for the means of existence disappears and is replaced by *co-operation*. He relates, for example, how a troop of baboons, threatened by hunters with firearms, rolled against them so many stones down the mountain that the hunters had to beat a hasty retreat. Darwin remarks that it "deserves notice that these baboons thus acted in concert."[2]

The fact is that throughout *The Descent of Man* Darwin brings into relief the fundamental role of interdependence and mutual aid in the adaptation and survival of species. Referring of the "social instincts," he argues that they were acquired by humans, as by the "lower" animals, for the good of the community, imparting to humans the inclination to aid their fellows and sympathize with them. Underlying the human moral constitution, the social instincts – with the aid of active intellectual powers and the effects of habit and experience – "naturally lead to the Golden Rule, "as ye would that men should do to you, do ye to them likewise,' and this lies at the foundation of morality."[3] If human beings have become the dominant animal on earth and have spread more widely than any other creature, they owe their dominance to their intellectual faculties and their social habits, which lead them to aid and defend their fellows. Furthermore, with strictly social animals, natural selection occasionally acts indirectly on the individual, through the preservation of variations which are beneficial only to the community. Darwin here addresses the paradox of the human being's relative weakness having led to his dominance. If man had descended from an animal possessing great size, strength, and ferocity which, like the gorilla, could easily defend itself against enemies, it is highly unlikely that man would have acquired his pronounced social qualities. Hence, writes Darwin,

> it might have been an immense advantage to man to have sprung from some comparatively weak creature. The slight corporeal strength of man, his little speed, his want of natural weapons, etc., are more than counterbalanced, firstly by his intellectual powers, through which he has, whilst still remaining in a barbarous state, formed for himself weapons, tools, etc., and secondly *by his social qualities which lead him to give aid to his fellow-men and to receive it in return.*[4]

Again and again Darwin underscores how central for human survival are the reasoning powers and foresight of group members, each learning that "if he aided his fellow-men, he would commonly receive aid in return."[5] Moreover, Darwin does not hesitate to attribute considerable causal influence to internalized moral experiences such as shame, guilt and remorse – and to human altruism. "It is incredible," he writes,

> that a savage, who will sacrifice his life rather than betray his tribe, . . . would not feel remorse in his inmost soul, though he might conceal it, if he had failed in his duty which he held sacred. It is highly probable that primeval man, at a very remote period, would have been influenced by the praise and blame of his fellows. It is obvious that the members of the same tribe would approve of conduct which appeared to them to be for the general good, and would reprobate that which appeared evil. To do good unto others – to do unto others as ye would they should do unto you – is the foundation-stone of morality. It is, therefore, hardly possible to exaggerate the importance during rude times of the love of praise and the dread of blame. A man who was not impelled by any deep, instinctive feeling to sacrifice his life for the good of others, yet was roused to such actions by a sense of glory, would by his example excite the same wish for glory in other men, and would strengthen by exercise the noble feeling of admiration. He might thus do far more good to his tribe than by begetting offspring with a tendency to inherit his own high character. (p. 165)

Darwin's studies of animals and humans therefore leave no doubt that interdependence, co-operation and mutual sympathy are essential if the species in question is to thrive and flourish. Nietzsche, however, somehow imagined that in the "struggle for survival" the outcome is the reverse of that proposed by the school of Darwin. For Nietzsche, Darwin's theory implied "the defeat of the stronger, the more privileged, the fortunate exceptions. Species do *not* grow more perfect: the weaker repeatedly dominate the stronger – the reason being they are the great majority . . ." (*Twilight of the Idols*, section 14). This conception of things flies in the face of the best scientific evidence in both Darwin's work and in post-Darwinian animal research. Under the heading of "Sexual Selection" in *The Descent of Man*, Darwin demonstrated that with social animals, the young males – who manifest their "will to power," in Nietzsche's terms – have to pass through many a contest before they win a female, and the older males have to retain their females by renewed battles. Darwin had dubbed this phenomenon, the "law of battle": "with animals in a state of nature, many characteristics proper to the males, such as size, strength, special weapons, courage, and pugnacity, have been acquired through the law of battle" (p. 371). And post-Darwinian research strongly suggests quite the opposite of what Nietzsche had imagined. There is neither any weakening of the group nor even any separation or antagonism between the "stronger type" and the herd; for the emergence of the stronger or victor benefits the herd by providing a leadership that heightens the effectiveness of the co-operative organization of the entire group. Field studies conducted over the past several decades on many different types of animals have shed considerable light on fighting between members of a species. The results of such studies have been summarized by the zoologist V. C. Wynne-Edwards, who has concluded that although

the stakes are sometimes life or death, serious fights and bloodshed are uncommon. Convention restricts the contestants very largely to displaying themselves for mutual appraisal or engaging in a harmless trial of strength, and from these actions they predict what the outcome of real combat would be without needing to fight it out. What they do is threaten or impress one another, at the crudest extreme by exposing or even briefly using their fighting weapons – butting with their horns or baring their teeth. In the most refined examples, the victor overrides the self-confidence of the loser by sheer magnificence and virtuosity.[6]

And as Adriaan Northland has observed,

> The goal of fighting in many species is not so much fighting in itself but
> rather to establish a social organization which makes fighting superfluous.[7]

In the same vein, J. L. Cloudsley-Thompson has maintained that

> threatening gestures and ceremonial displays frequently replace actual
> fighting. In this way conflict tends to become ritualized and adapted, so
> that its function may be achieved without harm to the rivals.[8]

Finally, as Ueli Nagel and Hans Kummer have succinctly stated,

> Aggression in animals is primarily a way of competition, not of
> destruction.[9]

In sum, the "aggressive" encounters between members of a species tend
to preserve peace within the group and contribute to the preservation of
the species. Whether it is a pack of wolves or any other group of animals,
the leader needs the followers no less than they need him.

If Nietzsche were confronted with the evidence considered here – that
peace, co-operation, mutual dependence, and the Golden Rule are the
implicit moral principles governing intra-species animal conduct in a
state of nature – one wonders whether he would accuse the animals of
somehow having been infected by the values of a "slave morality." In all
fairness to Nietzsche we should note that there is at least one aphorism of
his suggesting that he did in fact grasp the "moral" dimension of animal
conduct:

> The animal observes the effect it has on the perceptions of other animals,
> from which it learns to look back upon itself, to take itself "objectively";
> it has its degree of self-knowledge. The animal assesses the movements of
> its enemies and friends, it learns their peculiarities by heart, it prepares
> itself for them: it renounces war once and for all against individuals of a
> certain species, and it can also discern in the way certain kinds of animals
> approach, that they have peaceful and friendly intentions. The beginnings
> of justice, as of prudence, moderation, bravery – in short of all we denote
> as the *Socratic virtues*, are *animal*: a result of that drive which teaches us to

seek nourishment and to avoid enemies. If we now consider that even the highest human being has only become more elevated and discriminating in his choice of food and in his conception of what is hostile to him, then it is not improper to describe the moral phenomenon in its entirety as animal. (*Dawn*, section 26)

We see, then, that Nietzsche discerned the roots of values and virtues in the experiences of animals and in their "reflections" upon those experiences. Animals possess sufficient cognitive capacity and "practical intelligence" to determine what is good for them and what is not. This lesson from nature is, however, ignored throughout Nietzsche's work, where he exalts the "higher type" and denigrates the "herd," and looks with contempt upon the "slave morality." It seems clear, therefore, that if Nietzsche were forced to acknowledge in the light of the best evidence, that the intra-species behavior of animals largely coincides with what he calls "slave morality," he would no longer be able to deprecate that morality as "unnatural," sickly, decadent, and enervating.

NOTES

1 Charles Darwin, *On the Origin of Species by Means of Natural Selection* (John Murray, London, 1859), p. 62.
2 Charles Darwin, *The Descent of Man and Selection in Relation to Sex* (Princeton University Press, Princeton, N. J., 1981); photoreproduction of the 1871 edn (John Murray, London), p. 52.
3 Ibid., p. 106.
4 Ibid., pp. 156–7; italics added.
5 Ibid., p. 163. References to *The Descent of Man* are hereafter cited in parentheses immediately following the quoted passage.
6 V. C. Wynne-Edwards, "Ecology and the Evolution of Social Ethics," in J. W. S. Pringle, ed., *Biology and the Human Sciences* (Clarendon Press, Oxford, 1972), p. 59.
7 Adriaan Northland, "Aspects and Prospects of the Concept of Instinct," ibid., p. 207.
8 J. L. Cloudsley-Thompson, *Animal Conflict and Adaptation* (G. T. Foulis, London, 1965), p. 80.
9 Ueli Nagel and Hans Kummer, "Variations in Cercopithecoid Aggressive Behavior," in Ralph L. Halloway, ed., *Primate Aggression, Territoriality and Xenophobia* (Academic Press, New York, 1974), pp. 159–84.

10

The Twilight of the Idols

The subtitle of Nietzsche's book is: "or How to Philosophize with a Hammer." And, indeed, no thinker in the history of philosophy has wielded his hammer as unrelentingly as has Nietzsche. The first idol that Nietzsche sounds out with his hammer and finds rather hollow and fragile is no less a figure than Socrates. Earlier, in our discussion of Socrates and his participation in the Greek inversion of values, we observed that Nietzsche disliked Socrates for at least two reasons: his endorsement of the values and virtues associated with the "slave morality," and his belief in the actual existence of the transcendent, supradivine Forms and the eternity of the soul. Now, in the light of what Nietzsche has to say in this book concerning "The Problem of Socrates," we can say without hesitation that the very quality for which Socrates has been revered is the same quality for which Nietzsche detested him.

Nietzsche alleges that the wise in every age have judged life to be worthless and that Socrates, too, appears to have been weary of life, saying, as he died, "Crito, I ought to offer a cock to Asclepius. See to it, and don't forget to pay the debt" (*Phaedo*, 118A). As it was customary to offer a cock to Asclepius upon recovering from an illness, Nietzsche infers that Socrates was saying that life, for him, is or has been an illness. Here, in *The Twilight of the Idols* (1889), one of Nietzsche's last works, he tells us that it first dawned on him as early as *The Birth of Tragedy* (1872), contrary to the then prevalent view, that Socrates and Plato were symptoms and agents of the decay and dissolution of old Greece – the Greece of the noble-warrior age. Nietzsche begins his analysis by situat-

ing Socrates in the lower social strata. "Socrates was rabble," says
Nietzsche, which is evident from the fact that he was ugly – ugliness
itself having been among the Greeks a sure sign of the lack of nobility.
It is true, judging from Plato's description, that Socrates was far
from handsome, being snub-nosed and possessing protruding eyes and
walking with a peculiar gait which Aristophanes likened to the strut of
some sort of waterfowl. For Nietzsche, however, Socrates' "ugliness"
raises the question of whether he was a Greek at all. Why? Because,
he assures us, ugliness is frequently a sign of a thwarted or retarded
development caused by interbreeding. Citing the work of anthropologi-
cal criminologists apparently from the school of Cesare Lombroso,
Nietzsche proposes that the typical criminal is ugly – a monster in face,
a monster in soul. The criminal is a *décadent*! Nietzsche then asks: "Was
Socrates a typical criminal?" He replies in the affirmative, perhaps
tongue-in-cheek, but asserting nevertheless that Socrates contained
within him every kind of vice and lust.

Nietzsche appears to be altogether serious, however, when he attributes to Socrates an admitted "dissoluteness" and "anarchy of the
instincts" [*die zugestandne Wüstheit und Anarchie in den Instinkten*], a
sure indication of his decadence. Without citing the exact source,
Nietzsche refers to a famous contemporary physiognomist's opinion to
support the charge of dissoluteness (sections 3 and 4). The source seems
to be Cicero's *Tusculan Disputations* (IV. xxxvii. 79–81), where we read:

> Zopyrus, who claimed to discern every man's nature from his appearance,
> accused Socrates, in company, of a number of vices which he enumerated,
> and when he was ridiculed by the rest who said they failed to recognize
> such vices in Socrates, Socrates himself came to his rescue by saying that
> he was naturally inclined to the vices named, but had cast them out of him
> by the help of reason.

But this passage certainly does not support Nietzsche's allegation of
dissoluteness on the part of Socrates. On the contrary, the passage points
to a definite *self-mastery* which Socrates had achieved, the very quality
which Nietzsche supposedly admires. All that Socrates "admits" to in
the passage is to having had certain impulses, presumably of an erotic
sort, which he had successfully cast out of himself, that is, repressed

and sublimated. There is no evidence either here or in Plato that Socrates had ever engaged in *actions* that would warrant the accusation of dissoluteness.

In Plato's dialogue *Charmides* we hear a conversation between Socrates and Critias, the latter stating that Charmides "is as fair and good within, as he is without." Socrates then expresses his eagerness to meet this beautiful young man, so Critias calls Charmides and everyone pushes and shoves in an effort to sit next to him. When, finally, Charmides seats himself between Critias and Socrates, Plato attributes an extreme awkwardness to Socrates who, when he catches sight of Charmides' inner garment, becomes aroused – overcome by a "wild-beast appetite" (*Charmides*, 155C–E). At the end of the dialogue Charmides promises Critias that he will allow himself to be charmed by Socrates and never abandon him. Socrates asks what the two men are conspiring about, and Charmides replies that they are not conspiring, for they have conspired already. Socrates then asks whether Charmides intends to use force, and when he answers in the affirmative, Socrates states that if Charmides is in a violent and determined mood, he is irresistible (*Charmides*, 176C–D).

In the *Symposium* we hear the intoxicated Alcibiades berating Socrates for having sat next to Agathon, the handsomest man in the room. Socrates then asks Agathon, tongue-in-cheek, to protect him from Alcibiades who, he says, flies into a jealous rage whenever Socrates says a word to some other attractive young man or even merely looks at him. Hearing this, Alcibiades protests that it is the other way around and that it is Socrates who finds it intolerable when he hears Alcibiades say a good word about anyone else. Laughingly, however, Alcibiades then goes on to describe the several elaborate efforts he had made to seduce Socrates, all of which failed (*Symposium*, 213C–D, 219B–D). So the personal chastity of Socrates is assumed throughout. Socrates experienced strong erotic impulses toward certain young men but brought those impulses under control.

What also annoyed Nietzsche about Socrates were those auditory hallucinations which, as his "demon," were interpreted as a religious phenomenon (section 4). Nietzsche is referring here to Socrates' assertion that he had been divinely inspired (*Phaedrus*, 238C); it was well known that even as a boy he had a "voice" which he regarded as his divine sign. He believed he was sent to "this city [Athens] as a gift from the gods" and that he "was subject to a supernatural experience . . . It began in my early childhood – a sort of voice that comes to me" (*Apology*, 31C–D). Subject to ecstatic trances, he would stand motionless for hours buried in thought and quite forgetful of the outer world (*Symposium*, 220C–D). He was a firm believer in the immortality of the soul, as we observed earlier, and in rebirth and reminiscences from former lives. "Being immortal," says Socrates,

> the soul has been born myriad times and has witnessed all things both here and in the other world, and has thus learned everything. So we need not be astonished if the soul can recall the knowledge of virtue or anything else which, as we see, it once possessed. The things of nature are akin, and the soul has learned everything about them; so when an individual has recalled a single piece of knowledge there is no reason why he should not recover all the rest if he perseveres and does not tire of the search, for seeking knowledge and learning are, indeed, nothing but recollection. (*Meno*, 81B–D)

The soul strives for perfect purity, which it cannot achieve until it is released from the body. Then and only then can the pure soul dwell with the divine. For Socrates it is this mystical conception of the soul that defines the calling of the philosopher. Philosophy is the most sublime of activities, and the person who engages in it will therefore strive for "death" even in his lifetime by training the soul to concentrate upon itself and thus to obtain as much wisdom as possible in this world. Socrates, then, was a religious man, a visionary, an "enthusiast," a mystic. In the light of all this it is easy to see why Nietzsche the materialist would have detested this dimension of Socrates' outlook. But did Nietzsche admire him as the fount of reason and dialectics?

The answer to this question is a resounding "no," and the evidence is unequivocal (*Twilight* . . . , "The Problem of Socrates," sections 5–12). The Socratic equation that reason = virtue = happiness is not only the bizarrest of idiosyncrasies for Nietzsche, it has all the instincts of the older Hellenes against it. Nietzsche will now advance the dramatic proposition that reason and dialectics are one more product of the inversion of noble-warrior values. Reason is the invention of the slaves and the rabble. We are back to the thesis of Nietzsche's first book, *The Birth of*

Tragedy, a thesis extended and elaborated in *The Twilight of the Idols*: with Socrates, Greek taste underwent a change in favor of dialectics. What this means, for Nietzsche, is that the nobler taste was defeated and the rabble won out. Prior to Socrates, dialectical discourse was unacceptable in good society, that is, noble society. It was considered to be bad manners. Young people were warned against it, since giving reasons was regarded with mistrust, and honest men knew that one cheapens the value of one's words if one has to prove their validity. Wherever noble authority was still intact in old Greece, one did not give reasons, but simply commanded; and the dialectician, if he existed at all, was a kind of buffoon, laughed at and never taken seriously.

In the face of dialectics, Nietzsche contends, one is often distrustful and unconvinced. Notice how Socrates' dialectical prowess was often associated with that of the Sophists. One employs dialectics when one has no other expedient. Dialectics are a last resort, a last-ditch device in the hands of those who possess no other weapon. Reason is a weapon of the weak and impotent, a weapon which the powerful do not need since they simply enforce their will. If one claims certain rights and cannot enforce them, one cannot securely make use of those rights. That is why the Jews of Europe were dialecticians. "Reynard the Fox [*Reinecke Fuchs*] was a dialectician: What? and Socrates was a dialectician too?" (section 6). For Nietzsche, then, reason and dialectics are a product of the *ressentiment* of the rabble, a form of spiritual revenge. Socrates, emanating from the lower strata, employed the syllogism as a pitiless instrument with which to avenge himself on the aristocrats, whom he nevertheless fascinated. Why were they so fascinated by Socrates?

Nietzsche explains the fascination in which Socrates was held by the aristocrats by his perfection of a new kind of *agon* – a new kind of contest in which he was the first grandmaster. They were charmed by him because he touched upon the agonal instinct of the Hellenes; "he brought a variation into the boxing and wrestling matches [*in den Ringkampf*] between young men and youths" (section 8). But Nietzsche overlooks the fact that aristocratic young men found Socrates' skills *politically* valuable.

It is true that many young men from the privileged and oligarchically-inclined backgrounds were drawn to him, taking great pleasure in hearing the ignorance of others – Athenian politicians and plain people –

exposed. However, the ambitious aristocratic young men attached themselves to Socrates not merely for the amusement it entailed, but also to gain the dialectical skills necessary to compete for a successful political career in democratic Athens, a career in which they could defend and advance the interests of their class. They believed, evidently, that they could learn more from Socrates in that regard than from any professional Sophist. It is certain that Critias associated with Socrates for that reason. With the defeat of Athens in the Peloponnesian war, the extreme oligarchic party rejoiced in the foreign occupation, seeing it as an opportunity for the subversion of Athenian democracy. Of the exiles, the most prominent and determined was Critias, who had returned from exile bitter and revengeful toward the democracy and who became the mastermind behind the murderous oligarchic rule of the Thirty Tyrants imposed by Sparta.

Nietzsche, however, has no interest in this aspect of Socrates' relation to the aristocratic young men of Athens. Instead, his interest lies in the decline of old Athens which, he claims, was therefore ripe for the change Socrates had helped bring about. Everywhere, Nietzsche alleges, the instincts were in anarchy and the people were only a few steps from excess. If the instincts were thus threatening to become tyrannical, only reason could succeed in bringing them under control. Neither Socrates nor those who had come under his influence were free to be rational or not. Rationality was a last expedient. Reason became an essential element of the moralism of Greek philosophy, a pathologically conditioned element from the time of Plato downwards. The equation: reason = virtue = happiness meant that one must emulate Socrates by countering the dark desires through the light of reason. One must bring the desires to light, for every obedience to the instincts, to the unconscious, leads *downwards* (section 10). For Nietzsche, it is rationality together with the entire "morality of improvement" that constitutes decadence. Rationality stands in opposition to the instincts, and is therefore a sign of sickness, not health. In the combating of the instincts one descends into decadence; and in gratifying the instincts one ascends, for happiness and instinct are one. Socrates wanted to die, which means he must have been a long time sick. How could he have presumed to be a good physician when his will to power and life were so weak?

Not only is reason opposed to the instincts and to life; for Nietzsche, it has fundamentally misled philosophers for millennia, prompting them to create conceptual idols, which are simply impervious to change. "What is, does not *become*; what becomes *is* not," they have maintained. When it is objected that things do have the appearance of change, the philosophers reply: such appearances are an illusion, the work of the senses which deceive us about the real world. So have argued all those conceptual idolaters from the time of Parmenides (fifth century BC), who refused to acknowledge the evidence of the senses because they showed plurality and change. Heraclitus, in contrast, to whom Nietzsche accords "great reverence," recognized the eternal flux, though he, too, was unjust to the senses in failing to recognize that they do not lie. The lie is first introduced into sensory experience by "Reason," which posits unity, materiality, substance, and so on. "Reason" is the cause of the falsification of the evidence of the senses. Thus Nietzsche appears to make a radical break with dialectics and reason, for they are not only integral elements of the slave morality, they are the primary source of our basic epistemological error. Philosophers are bound to perpetuate the error as long as their prejudice in favor of reason compels them to posit unity, identity, duration, substance, cause, materiality, and so on (*Twilight . . .* , "Reason in Philosophy," section 5). Has Nietzsche thus reverted to a strict empiricism? We know from his frequent attacks upon Kant that he rejects the Kantian mediation between sensory experience and reason, and the notion of the "thing in itself" in particular. Characteristically, however, Nietzsche is elusive, and provides no affirmation with which to replace what he has attempted to tear down. He does, however, advance four propositions to make it easier, he says, to grasp where he stands epistemologically:

1   The apparent world (the one subject to sensory experience) is the only real one; no other kind of reality is demonstrable.
2   The characteristics which have been attributed to "real being" (e.g., the Socratic-Platonic Forms) are actually characteristics of non-being, of *nothingness*.
3   It is pointless to talk about "another" world. Only those whose will to life is weak avenge themselves on life by means of the phantom of another or "better" life.

4   To divide the world into a "real" and "apparent" one, as did
    Christianity and Kant, is a sign of decadence and declining life. That
    the artist values appearance more than reality is no objection to this
    thesis, since "appearance" in that case is in fact reality, only selected,
    accentuated, and corrected. "The tragic artist is *no* pessimist – it is he
    who says *yes* to all that is questionable and terrible in existence, he is
    Dionysian . . ." (section 6).

It is clear that Nietzsche rejects and repudiates both metaphysics and
reason, the latter also being a metaphysic. And what he affirms is the
body, the instincts and the senses, and the acknowledgement of all that is
"questionable and terrible in existence." So we are back to Dionysus, and
the question is: where does that leave us? In *The Birth of Tragedy*
Dionysus stood for the powerful, passionate element in art and life, while
Apollo represented the form-creating principle. The greatness of
Greek tragedy is conceived as the result of Apollo's harnessing and
directing of the Dionysian passions. In a later state of Nietzsche's
thinking this mythological-metaphorical idea is transformed into a psychological one, in which the Apollinian harnessing of the Dionysian
passions and instincts represents the self-overcoming of the "animal"
man. "Apollinian" implies that something is harmonious, measured,
ordered or balanced in character. If these are purely aesthetic concepts
for Nietzsche, then it becomes strictly a matter of *taste* whether a way of
life or a work of art is harmonious, ordered or balanced.

From *Zarathustra* on, the Apollinian (and the Socratic) element disappears entirely, and in place of the older duality Nietzsche now
recognizes only one force in the human constitution, the "will to power."
There are scholars who contend that "Dionysus," which formerly stood
for the unbridled passions, now presumably stands for the sublimated
will to power and is therefore synonymous with the *Übermensch*, the
superman in whom the will to power is sublimated into creativity. From
this standpoint, Nietzsche's war-metaphors, his admiration for the
"master morality," his esteem for historical persons like Alcibiades,
Julius Caesar, Cesare Borgia and Napoleon, and, finally, his proclamation of the superman, are all to be understood as the sublimated and
spiritualized symbols of the creative powers of the highest specimen of
artist. It is almost certain that Nietzsche would have accepted the

characterization of his entire *oeuvre* as a series of thought-experiments uttered from an aesthetic point of view. But this only strengthens our thesis that Nietzsche was an immoralist – the reason being that for Nietzsche, in life as in art, neither morality nor reason may be admitted as criteria for the assessment of a "creative" act. Hence, "taste" remains the sole criterion for the evaluation of human actions and relationships; and the "superman" or "higher types" become the sole arbiters of taste.

Moreover, whether we look at Nietzsche's work as a theory of art or as a general philosophy of life, an unsolved problem becomes evident with Nietzsche's abandonment of the Dionysian-Apollinian dualism: if Dionysus stands alone in the post-*Zarathustra* writings, where is the harnessing element? Indeed, if the element of reason has been entirely expunged from the creative process, the problem remains even if the Apollinian element is present. This would be true because Nietzsche conceives of both antithetical concepts, Apollinian and Dionysian, as forms of *intoxication*. In Apollinian intoxication it is primarily the eye that is alerted so that it acquires heightened powers of vision. Painters, sculptors and epic poets are examples of such visionaries. In Dionysian intoxication, on the other hand, the entire emotional makeup of the individual is alerted and elevated, so that it vents all its powers of representation, imitation, transfiguration, play-acting and so forth. The Dionysian man recognizes every signal from the emotions, "he possesses to the highest degree the instinct for understanding and divining, just as he commands the communicating art to the highest degree" (*Twilight* . . . , "Expeditions of an Untimely Man," section 10).

What we have here, then, are two forms of intoxication, one which alerts and intensifies the powers of the eye, the other which does the same for the entire emotional state. Sensory experience and instinct, therefore, are, for Nietzsche, the sole elements in the creative process. This has the ring of truth where the plastic arts, music and dance are concerned; for Nietzsche would no doubt argue that knowledge and reason are required at the technical level, but not beyond that, not in the creative process itself. But even if we accept the validity of Nietzsche's theory as applied to art, it is nevertheless certain that the senses and the instincts are insufficient for guidance in social life. Yet Nietzsche persists throughout in viewing virtually every form of morality as if it were nothing more than a loud, impudent condemnation of life, nature and the instincts.

And what is the nature of his "affirmations"? The truth is that they amount to neither more nor less than envisioning an age of "higher types" whose beyond-good-and-evil will to power and instincts will supersede morality. Insofar as Nietzsche will allow us to speak of good and bad at all, it is in these terms:

> Every error, of whatever type, is a result of the degeneration of instinct and vitiation [*Disgregation*] of the will: one has thereby nearly defined the *bad*. Everything *good* is instinct . . . (*Twilight* . . . , "The Four Great Errors," section 2)

One must not confuse the "will to power" with the notion of "free will," the latter being an egregious error for Nietzsche. The doctrine of free will was invented for the purpose of attributing guilt and making people accountable for their acts. The error lies in the assumption that one is impelled to act by one's reason and consciousness. The "will to power," in contrast, is rooted in the instincts. Nietzsche seems not bothered by the fact that instincts are, by nature, blind. In fact, he seems to believe in what the Greeks called *moira* (fate), a supradivine, impersonal force. No one is accountable, says Nietzsche, either for being constituted as he is, or for the circumstances in which he finds himself. In an implicit allusion to his notion of the "eternal recurrence," he asserts the one cannot disentangle the fatality of one's nature from the fatality of all that has gone before and all that will be. One is simply a necessary piece of fate, belonging in the whole and being in the whole. As nothing exists apart from or outside the whole, nothing exists by which to judge, measure, compare, or condemn our being. In this way Nietzsche proposes that everything in the universe is subject to a blind, necessary, impersonal force. "Free will" is a chimera. Thus Nietzsche believes in the existence of *fate*. Is this belief any less metaphysical than the Greek belief in *moira*? Furthermore, if the great chain of being is inexorably cyclical – as is implied in the notion of "eternal recurrence" – what is the point of proclaiming and waiting for the superman and railing against the "improvers" of mankind who are actually its "tamers" and "deteriorators"?

Like Stirner, Nietzsche values egoism and despises altruism as a decadent morality under which the ego languishes. In response to harsh criticism of his concept, "beyond good and evil," and his selection of Cesare Borgia as an example of a strong-willed "higher man," Nietzsche asks this question: has the world grown more moral? In the asking of that question he wishes, of course, to make the weighty point that all the "improvers" of mankind together with their moral doctrines may have utterly failed to make a difference. However, had he lived to see the amoral, totalitarian regimes of the twentieth century, he might have changed his opinion; for each and every one of them promoted "beyond-good-and-evil" ideologies of one kind or another. Moreover, Nietzsche's rhetoric may have contributed something to the formation of such ideologies. In clarifying his conception of freedom, for instance, he writes:

> war educates and trains for freedom. For what is freedom? That one has the will to self-responsibility. That one holds firm the distance which separates us. That one has become more indifferent to hardship, toil, deprivation, even to life. That one is prepared to sacrifice people, oneself not excepted, to one's cause. Freedom means that the manly, war- and victory instincts have gained mastery over other instincts, over the instinct for "happiness," for example. The man *who has become free*, and how much more the *spirit* that has become free, treads with contempt upon the sort of well-being dreamed of by shopkeepers, Christians, cows, women, Englishmen and other democrats. The free man is a *warrior*. (*Twilight . . .* , "Expeditions of an Untimely Man," section 38)

And in the same aphorism:

> The nations which were worth something, which *became* worth something, never became so under liberal institutions: it was *great danger* which made of them nations deserving reverence; [it is] danger which first teaches us to know our resources, our virtues, our weapons for war [*unsre Wehr und Waffen*], our *spirit* – which forces us to be strong. (section 38)

It is perhaps possible to "spiritualize" this language and to interpret such passages to the effect that the highest type of free individual is one who courageously confronts and overcomes resistance and danger. Psychologically this requires the successful combating of all impulses that militate against the maximum discipline over oneself, of which

Julius Caesar, for Nietzsche, is the finest example. But although we can agree that self-mastery is a virtue, Nietzsche's conception of it is egoistic in the extreme: he is exclusively concerned with *self*-responsibility, and thus never with responsibility to others. It is therefore easy to see how such passages together with Nietzsche's amoral glorification of power could serve as the ideological justification for the "master race" and other Nazi tenets. Short of that, one can say there are many other assertions which, being unsupported, amount to no more than an enunciation of Nietzsche's own snobbish and élitist taste in values: one cannot establish marriage and the family on the basis of "love," he asserts; one can only establish it on the basis of the "*drive to dominate*" (section 39). On the labor question, workers have been given the vote and the right to form unions: "if one wants slaves, one is a fool if one educates them to be masters" (section 40). "The doctrine of equality! . . . But there exists no more poisonous poison" (section 48). Finally, we need to observe that even when Nietzsche's example of the higher type is not a military figure but rather Goethe, Nietzsche gives him special license: the higher type is "a man to whom nothing is forbidden" (section 49). As we shall see in our discussion of Dostoevsky's Challenge, the idea that nothing is forbidden to certain types of men was bound to be appropriated and put into practice in a manner that might have terrified even Nietzsche.

11

The Anti-Christ

In *The Anti-Christ*, Nietzsche analyzes the New Testament in consider-able detail, explaining more clearly than in any of his previous works the reason for his animus towards Christianity. The reason for his hostility toward the New Testament and the religion to which it gave rise is similar to the reason for his ill will towards the "morality-and-ideal swindle" of the Socratic schools. Nietzsche despises Platonism because it represents the *décadence* (he always uses the French word) of the Greek instinct. For Nietzsche, the leading Athenian intellectual of the time was neither Socrates nor Plato, but Thucydides, who never deceived either himself or others in dealing with reality. In Nietzsche's description of Thucydides we hear the most succinct characterization of the "higher type" intellectual and its antipode. Thucydides was

> the grand summation, the last manifestation of that strong, rigorous, hard realism instinctive to the older Hellenes. *Courage* in the face of reality ultimately distinguishes such natures as Thucydides and Plato: Plato is a coward in the face of reality – consequently he flees into the ideal; Thucydides has *himself* under control, consequently he maintains control over things . . . (*Twilight*, "What I owe to the Ancients," section 2)

As is the case with Platonism, the fundamental defect of the Christian world of ideas is that it is cowardly when confronted with reality and thus flees into the ideal. To understand Nietzsche's standpoint in his bitter "higher criticism" of the New Testament, we need to review his critique rather carefully.

For Nietzsche, the word "Christianity" is misleading because there was in reality only one Christian, and he died on the cross. It is false and hypocritical to posit faith in redemption through Christ as the distinguishing feature of a Christian, since only those who, in practice, live a life such as the one Jesus of Nazareth lived deserve to be called Christians. Being Christian is not merely a matter of belief or a state of consciousness; it is a matter of doing and *not*-doing certain things. Have there been any true Christians during the past two millennia? How could there have been, when the Church has always spoken of *faith* and has presented to its adherents a world of ideas containing nothing which so much as touches upon *reality*? Christ lived and died in an exemplary way as a demonstration of his teaching. His disciples, however, transformed his death into an affair requiring "retribution," "punishment," "revenge," and "judgment" – products of a bitter *ressentiment*.

For Nietzsche, it was primarily Paul who was responsible for beginning the transformation. With Paul the real meaning of Jesus' exemplary life and death was totally obliterated and replaced with a "horrible paganism" in which God gave his son, an innocent man, for the sins of the guilty. Neither the reality nor the historical truth became the foundation of the new religion, but a falsified history in which the entire Old Testament drama of Israel was interpreted as the pre-history of Christianity. Paul shifted the meaning of Christ's life and death from the real world to the beyond – to the "*lie* of the 'resurrected' Jesus" (section 42). Thus ignoring the redeemer's *life*, Paul made of his hallucination on the road to Damascus the "proof" that the redeemer was still alive. Nietzsche even doubts that Paul had such an hallucination, and suspects that in his "priestly" quest for power Paul simply devised new teachings as a means of creating a spiritual tyranny over the masses.

The original and actual meaning of Jesus' life was thus shifted away from life and into the "beyond," into nothingness. This was accomplished by means of the big lie of personal immortality, a doctrine contrary to reason, which opposed all the instincts for life. Life after death became the "meaning" of life. Christianity thus owed its victory to the doctrine of the immortal soul, in which everyone is equal to everyone else, and in which the salvation of every single individual may claim to be of everlasting moment. In thus appealing to egoism and personal vanity, Christianity was able to win over to its side the underprivileged and rebellious-minded masses who wanted to believe that the "world revolves around *me*" (section 43). This is the source of that poisonous "democratic" doctrine of equal rights for all. From the time of Paul, Christianity has waged a war against the social distances between man and man, thus forging out of the *ressentiment* of the masses its chief weapon against everything noble, joyful, and high-spirited on earth, against happiness in the one and only world that exists. The Christian doctrine of the "equality of souls" is ultimately responsible for the modern, sickly state of affairs in which no one any longer has the courage to claim special privileges or the right to rule. The aristocratic outlook (the "master morality") has been permanently undermined by the Christian revolt against everything elevated. The Gospel of the humble and lowly makes everyone low.

Nietzsche detests the Gospels for their pretense of holiness. They say, "Judge not!" but they send to hell everything that stands in their way. In glorifying God, they actually glorify themselves. They have appropriated "morality" as a means of seducing mankind and leading it by the nose. What poses as modesty is in reality the arrogance and megalomania of the self-chosen *Electi*, the so-called "good and just" who are on the side of "truth," and who relegate everyone else to the other side. Nietzsche provides examples of the vengeful inversion of noble values one finds in the Gospels and in Paul's letters, examples of what the petty authors had placed in the mouth of their Master. The passages in question are quoted and then followed by Nietzsche's sardonic comments:

> *Mark 6:11*　And if any place will not receive you and they refuse to hear you, when you leave, shake off the dust that is on your feet for a testimony against them. "How evangelic!"

> *Mark 9:42*　Whoever causes one of these little ones who believe in me to sin, it would be better for him if a great millstone were hung around his neck and he were thrown into the sea. "How evangelic!"

> *Mark 9:47–48*　And if your eye causes you to sin, pluck it out; it is better for you to enter the Kingdom of God with one eye than with two eyes to be thrown into hell. "It is not precisely the eye that is meant . . ."

Mark 9:1 Truly, I say to you, there are some standing here who will not taste death before they see that the Kingdom of God has come with power. "Well lied . . ."

Matthew 7:1–2 Judge not, that you be rot judged. For with the judgment you pronounce you will be judged, and the measure you give will be the measure you get. "What a conception of justice, of a 'just' judge! . . ."

Matthew 5:46–47 For if you love those who love you, what reward have you? Do not even the tax collectors do the same? And if you salute only your brethren, what more are you doing than others? Do not even the Gentiles do the same? "Principle of 'Christian love': it wants to be well *paid* . . ."

Matthew 6:15 But if you do not forgive men their trespasses, neither will your Father forgive your trespasses. "Very compromising for the said 'Father' . . ."

Matthew 6:33 But seek first his Kingdom and his righteousness, and all these things shall be yours as well. "All these things: namely food, clothing, all the necessities of life. An *error*, to put it mildly . . ."

Luke 6:23 Rejoice on that day, and leap for joy, for behold, your reward is great in heaven; for so their fathers did to the prophets. "Impudent rabble! It already compares itself with the prophets . . ."

I Corinthians 3:16–17 Do you not know that you are God's temple and that God's Spirit dwells in you? If anyone destroys God's temple, God will destroy him. For God's temple is holy, and that temple you are. "Things like this one cannot sufficiently despise . . ."

I Corinthians 6:2 Do you not know that the saints will judge the world? And if the world is to be judged by you, are you incompetent to try trivial cases? "Unfortunately, not merely the ravings of a lunatic. . . . This *frightful impostor* goes on to say: Do you not know that we are to judge angels? How much more matters pertaining to this life!"

I Corinthians 1:20f. . . . Not many wise men after the flesh, not many mighty, not many noble, are called: but God has chosen the foolish things of the world to confound the wise; and God has chosen the weak things of the world to confound the things that are mighty; and base things of the world, and things which are despised, has

God chosen, yea, and things which are not, to bring to nought
things that are: that no flesh should glory in his presence (see *Anti-
Christ*, section 45).

Nietzsche comments on this last passage that in order to understand it
one must read the first essay of his *Genealogy of Morals*, where he first
illuminated the opposition between a noble morality and a Chandala (i.e.,
outcast or rabble) morality born of *ressentiment* and powerless vengefulness. "Paul," he says, "was the greatest of all apostles of revenge . . ."
(section 45).

In Nietzsche's view, then, there are in the New Testament only
"bad," that is, contemptible, traits. Everything in it is cowardice and
self-deception in the face of life. When given opponents like Paul and the
Gospel authors, the Scribes and Pharisees gain advantage. Hated in such
an intense and indecent fashion, they must have been worth something.
Why were they hated? Was it for their hypocrisy? No, argues Nietzsche,
it was their *privileged* status that evoked so intense a reaction. The "first
Christians" were rebels against everything privileged – against the noble
values of intellectual honesty, manliness, pride, and integrity. For
Nietzsche, of course, there is no God either in history, in nature, or
above nature; but his primary aim in *The Anti-Christ* is not to deny the
existence of God, but to propose that the Christian conception of God is
a harmful illusion, a "crime against life." And if the existence of such a
God were proved to exist, he would believe in him even less. "God," as
Paul created him, is a denial of God, since his doctrine of faith gave birth
to a religion which at no point comes in contact with actuality.

Nietzsche sees the origin of the fear and denigration of knowledge in
the opening chapters of *Genesis*. God creates man and other animals
whom man finds boring. So God creates woman and "then, indeed, there
was an end to boredom"; but it was through her that man learned to taste
of the tree of knowledge. This deed struck fear into God, says Nietzsche,
so he drove man out of the garden of Eden lest he become God's rival.
The moral: knowledge is forbidden! Man shall not look around him; he
shall not look into things in order to learn; he shall not learn at all.
Instead, he shall *suffer*, and because of his suffering he shall have need of
the priest and of a Saviour. All the concepts associated with Faith – sin,
guilt, punishment, grace, forgiveness, redemption – are lies which were

created to destroy the causal sense in human beings. The Church has thus inculcated an *idée fixe* which turns human beings into weak and sickly creatures who fear their own bodies and combat instinctual health as a satanic temptation. The "holiness" with which the people are indoctrinated impoverishes, enervates, and corrupts the body. Nietzsche dissents from the prevalent scholarly opinion that the corruption of noble antiquity made Christianity possible. The opposite is true; the nobility of the Roman empire existed at the time in its highest and maturest form. It was the "democratization" of the Christian teachings that enabled them to conquer. Christianity was victorious because it appealed so strongly to the disinherited of every kind and thus found allies everywhere. Or as Paul had expressed it, God has chosen the weak, the base and the despised.

"Faith," for Nietzsche, means not wanting to know what is true: every frank, honest, scientific pursuit of knowledge has been prohibited by the Church. Even to doubt is a sin. The existence of martyrs has somehow been taken as proof of the truth of the martyr's cause. Martyrdom has led to epidemics of death-seeking, but as Zarathustra declared:

> Letters of blood they wrote on the path they followed, and their foolish-ness taught that with blood one proves the truth. But blood is the worst witness of truth; blood poisons the purest teaching [and turns it into] delusion and hatred of the heart. And if someone goes through fire for his teaching – what does that prove? It is more true when one's own teaching comes out of one's own burning. (*Thus Spoke Zarathustra*, Part Three, "Of the Priests," quoted by Nietzsche in *Anti-Christ*, section 53)

"Faith" and "convictions" are blinders. They are designed to prevent one from seeing certain things. They compel one to be partisan through and through, and to view all values from one narrow perspective. The man of faith and conviction is therefore the antagonist of the man of truth. And yet the conspicuous men of conviction – Savonarola, Luther, Rousseau, Robespierre, Saint-Simon – have impressed the masses, who would rather listen to fanatics than to reason. A conviction is, in effect, a lie; for what is a lie if not a refusal to see what the eye actually sees? Self-deception is, in that regard, more common than the deception of others. Nietzsche here widens his line of attack to include scholars, philosophers, and founders of other religions. Kant, who is often Nietzsche's favorite target, is derided for his metaphysical view that there are questions whose truth or untruth *cannot* be decided by human beings. For Kant, the supreme questions of value are all beyond human reason. One must acknowledge the limits of reason and recognize divine revelation as the source of our knowledge of good and evil. Consciously or not, Kant had thus furthered the ideological conviction that the priest is the only reliable authority for what is right and what is wrong. The "Law," the "Will of God," the "Sacred Book," "Revelation" – all of these are the means by which the priest comes to power and maintains it. All priestly-philosophical power-structures rest on such means. Hence, the "holy lie" is common to Confucius, Mohammed and the Christian Church. It is found in Plato, too, who made the truth a self-subsistent, metaphysical entity existing above the gods. In the Law-Book of Manu, in contrast, one breathes a refreshing atmosphere. It is fundamentally different from any sort of Bible in that the laws are the means

> by which the noble orders, the philosophers and the warriors, keep the multitude under control; noble values everywhere, a feeling of perfection, an affirmation of life, a triumphant feeling of well-being in oneself and towards life – the *sun* shines on the entire book. (*Anti-Christ*, section 56)

We will notice in this passage that it is the philosophers and warriors who keep the multitude under control. This bears a remarkable resemblance to Plato's three-class scheme in the *Republic*, except that Plato's philosopher-kings do not share the full affirmation of life which Nietzsche admires, since Plato deprives them of both family and property, and subjects them to an exceedingly ascetic way of life. For all his detestation of Plato, Nietzsche nevertheless adopts his three-class system, claiming that such a system is inherent and natural in any healthy human society. Nature, he says, separates from one another the predominantly spiritual, muscular, and mediocre types, the last being the majority and the first the ruling élite. As is the case in Plato's *Republic*, Nietzsche's highest caste also consists of the very few and as such possesses the privileges of the very few, which Nietzsche describes solely in the abstract terms of happiness, beauty, and benevolence on earth. But lo and behold, as Nietzsche proceeds to describe the highest, spiritual caste, he, too, subjects them to severe "self-constraint": with them "asceticism becomes nature, need, instinct" (*Anti-Christ*, section 57). They rule because of who they are, and their pursuit of knowledge is itself a form of asceticism; though, Nietzsche adds, that does not prevent them from being most cheerful and amiable. The second caste consists of the noble warriors who are the guardians of law and order. Finally, there is the majority of the populace, those who engage in agriculture, crafts, and trade, and who are in no way capable of anything other than mediocrity in ability and desires. This three-class system is therefore virtually identical with that of Plato's *Republic*. And while Plato envisions this system as an ideal, Nietzsche asserts that the order of rank and castes he has prescribed follows the supreme law of nature and life itself. The separation of ranks is not only necessary for the preservation of society, but also for enabling higher and higher types to emerge. "*Inequality* of rights," he declares, "is a precondition for the general existence of rights" (section 57). The existence of the mediocre mass is a prerequisite for the rise of the higher specimens. A high culture is conditional upon it. Nietzsche states that specialization is for the mediocre a natural instinct, bringing them a kind of happiness that comes from serving as a cog and fulfilling a single function. Nietzsche failed to realize, perhaps, that his entire caste system rests on specialization: the spiritual élite specializes in knowledge, the second in war, and the third in production. So in this respect Nietzsche is a mere footnote to Plato, replicating his scheme in detail. Plato was no democrat, and of course neither was Nietzsche, who sees democracy, socialism, and anarchism as emanating from the same psychological roots as Christianity, namely, from weakness, envy and revengefulness (section 57) – which brings us back to Paul.

It was Paul who divined that the small sectarian movement on the edge of Judaism could ignite a "world conflagration." By means of the symbol of "God on the cross" he could mobilize all the downtrodden and all those in secret revolt, and form them into a formidable power. Paul of Tarsus, with his sophisticated Hellenistic background, could borrow themes, perhaps unconsciously, from the other subterranean cults – of Osiris, of the Great Mother, of Mithras, for instance – and place them in the mouth of the "Saviour" he invented, so as to make of him something which the non-Jewish world could readily understand. Christianity thus robbed the West of the high culture of the ancient world, and then went on to rob the West of what the Islamic Moorish culture had preserved and revived from the ancient legacy. With the Renaissance a last effort was made, with artistic and intellectual genius of every kind and with every political expedient, to invert Christian values and to restore to supremacy the noble values. The attempt failed, however, due to the efforts of that German monk, Luther, who went to Rome and fulminated there against the Renaissance. "Instead of grasping with deep gratitude the colossal event which had occurred," writes Nietzsche,

> the overcoming of Christianity in its very seat, his [Luther's] hatred knew only how to draw nourishment from this spectacle. Luther saw the *corruption* of the Papacy, while precisely the opposite was tangibly obvious: the old corruption, the *peccatum originale*, Christianity no longer sat on the papal throne. On the contrary, Life sat there! The triumph of life! The great Yes to all elevated, beautiful, bold things! (section 61)

Luther revived Christianity, creating a new kind, Protestantism, which is the hardest to refute. Thus Nietzsche concludes that Christianity is a great curse upon mankind which has drawn its nourishment from human distress and which has perpetuated the mental suffering of its subjects in order to eternalize itself.

Did Nietzsche ensnare himself in a contradiction when he advanced the claim that a morality must be judged by its ability to foster the growth of the powerful? The corollary of that claim is that any morality which elevated the weak to positions of honor was harmful because it thwarted the aims of the naturally superior and powerful. Now, according to Nietzsche, as we have seen, Christian morality was among the most debilitating. Yet Christian morality was the officially ruling morality of his day. Does this mean that the weak had triumphed over the strong? Or, since Christianity had triumphed, might that not mean that it was actually the morality of the stronger? On what ground, then, could Nietzsche repudiate Christianity as a morality of weakness? Furthermore, if, as Nietzsche repeatedly insists, the only criterion of the "good" is power, must not every victorious force be called "good" simply because it has become victorious? Doesn't that reduce itself to admiration for "success," whatever its form and cost in human life and suffering? This is the fundamental defect of Nietzsche's philosophy which fails to recognize the need for universal ethical principles by which to distinguish between good and evil. In this light Nietzsche has failed, just as every philosophy must fail if it attempts to go "beyond good and evil."

12

Dostoevsky's Challenge

In *The Twilight of the Idols*, in his discussion of the "criminal," Nietzsche remarks:

> In regard to the problem before us the testimony of Dostoevsky is of importance – Dostoevsky, the only psychologist, by the way, from whom I had anything to learn: he is one of the happiest accidents of my life . . . ("Expeditions of an Untimely Man," section 45)

In a letter of 23 February, 1887, he wrote:

> a few weeks ago I did not even know the name Dostoevsky . . . An accidental reach in a bookshop brought before my eyes a book just translated into French, *L' Esprit souterrain* [*Notes from Underground*] . . . The instinct of kinship (or what shall I call it) spoke immediately, my joy was extraordinary . . . (*Chronik zu Nietzsches Leben*, vol. 15, p. 163)

By the seventh of March of the same year he had also read *The House of the Dead* and *The Insulted and The Injured*, both also in French translation. As for Dostoevsky's four great novels, we have direct evidence of his having read only *The Possessed*, since he discusses two of its main characters, Kiriloff and Stavrogin (*Nachgelassene Fragmente 1887–1889*, vol. 13, pp. 142 and 144), but makes no mention of the three other novels in his voluminous correspondence of 1887–8.

CRIME AND PUNISHMENT (1865–6)

In this extraordinary work, the most exciting and philosophically important crime novel ever written, Dostoevsky (1821–81) creates an unforgettable character named Raskolnikov, the first of Dostoevsky's characters in whom the problems of crime and value intertwine. Raskolnikov, a young impoverished university student, has been smitten by the ideas of the Enlightenment and, in particular, by the idea that "God is dead." He therefore flaunts a "beyond good and evil" attitude. He is a would-be "superman" who wants to demonstrate the strength of his personality by disregarding the values of good and evil and thus proving to himself that he is a higher type, a law-giver, a new Napoleon.

Raskolnikov has pawned his ring with a miserly old woman money-lender named Alyona Ivanova, and though he knows nothing about her, conceives a violent aversion for her. On his way home a strange but fascinating idea begins to hatch in his brain. He stops at a small restaurant, orders some tea, sits down, and falls into thought. Quite by coincidence, he finds himself within earshot of another table where a student chats with an army officer, telling him all sorts of details about the same Alyona Ivanova – that she is very rich and that she lends thousands, but is also bad-tempered. If one is even one day late in redeeming a pledge, one is likely never to see it again. She lends only about a quarter of what the article is worth and charges between five and seven percent a month. The student also mentions that the old woman moneylender has a sister, Lisaveta, whom she bullies and beats continually. The student then adds that he would gladly murder that damn old woman and rob her of all she has. Raskolnikov, listening intently to the conversation, then hears the student ask the officer a serious question: on the one hand, we have a miserly, wicked, worthless, decrepit old hag who is of no use to anyone and who will soon be dead anyway; on the other, we have large numbers of young and promising people who are going to rack and ruin without anyone lifting a finger to help them. Hundreds and maybe thousands of them could be saved, rescued from poverty, decay, and ruin with her money. Why not kill her, take her money, and with it devote yourself to the service of humanity? Don't you think that one such little crime could be expiated by thousands of good deeds? And come to think of it, what does the life of a sickly, wicked old hag amount to when weighed in the scales of the general good of mankind? It amounts to no more than the life of a louse or a black beetle. The army officer then stops the student to ask a question of his own: "Would you kill the old woman yourself?" To which the student replies, "Of course not! I was merely discussing the issue from the standpoint of justice; but personally I'd have nothing to do with it." Hearing all this, Raskolnikov becomes greatly agitated. Why had this discussion taken place at precisely the moment when he is turning over the same ideas in his own mind? It strikes him that this was, perhaps, a *sign*, and that there is something preordained here.

As the plot began to take shape in Raskolnikov's head, he anxiously reflected on something he had heard somewhere, to the effect that every criminal is subject at the moment of the crime to a kind of breakdown of his reasoning faculties and willpower, which are replaced by childish carelessness just at the moment when he is most in need of caution and cold reason. But Raskolnikov soon allays his anxiety by persuading himself that what he is planning to do is not really a crime. On the pretext of bringing the old woman the pledge he had promised on his earlier visit, Raskolnikov gains entry to her apartment, takes out the hatchet with which he has armed himself and almost mechanically crushes her skull with the back of it; there is no doubt that she is dead. He then runs frantically around her bedroom, rummaging through her clothes and soon finds a large variety of gold articles – all pledges no doubt – and wastes no time in stuffing them into the pockets of his trousers and coat. Just then, however, he hears footsteps in the apartment and as he rushes out of the bedroom, hatchet in hand, he sees, in the middle of the room, Lisaveta, the old woman's sister, looking petrified at the dead body of her sister. He lunges at her with the hatchet and the blow falls straight across her skull. He is now more and more seized with panic, especially after this second, quite unexpected murder.

Thus Raskolnikov, to prove to himself that he is a higher specimen, chooses murder as an expedient – the very crime by which one human being can assert a maximum of will and power over another human being. But Raskolnikov happens also to be a new variety of the brooding "underworld man." He dreams of becoming a Napoleon, but is not quite sure he is entitled to cherish such ambitions. In a moment of weakness he even says to a little girl, "My name is Rodion; please say a prayer for me too, sometimes." But he soon shakes off his doubts and resolutely says to himself, enough of these imaginary terrors.

"There is such a thing as life! Life is real! Haven't I lived just now? My life hasn't come to an end with the death of the old woman! . . . Now begins the reign of reason and light – and of will and strength – and we'll see now! We'll try our strength now," he added arrogantly, as though addressing some dark power and challenging it. . . . "What I want is strength – strength! You can't get anything without strength, and strength must be won by strength . . .

But you asked her to mention "thy servant Rodion" in her prayers!, it flashed through his mind suddenly. "Well" he added, "that was – just in case!" and he at once burst out laughing . . .[1]

In the course of a conversation later in the novel, Porfiry mentions an article in which Raskolnikov anticipated by several years the main thesis of Nietzsche, that mankind is divided by nature into two categories: an inferior one, whose only purpose is to reproduce its kind, and a superior one that possesses the gift or talent to say a new word. The first category, the masses, comprises all those who are ordinary and docile and love to be docile. In contrast,

the men belonging to the second category all transgress the law and are all destroyers, or are inclined to be destroyers, according to their different capacities. The crimes of these people are, of course, relative and various; mostly, however, they demand in proclamations of one kind or another the destruction of the present in the name of a better future. But if for the sake of his idea such a man has to step over a corpse or wade through blood, he is, in my opinion, absolutely entitled, in accordance with the dictates of his conscience, to permit himself to wade through blood, all depending of course on the nature and scale of his idea . . . (277)

The great mass of the people – the masses – exist merely for the sake of bringing into the world by some supreme effort, by some mysterious process we know nothing about . . . one man out of a thousand who is to some extent independent. (279–80)

Raskolnikov wants to think of himself as one of those extraordinary, independent specimens, but he is plagued by bouts of self-doubt which he strives to overcome by identifying with Napoleon:

a real *ruler of men*, a man *to whom everything is permitted*, takes Toulon by storm, carries out a massacre in Paris, *forgets* an army in Egypt, *wastes* half a million men in his Moscow campaign, and gets away with a pun in Vilna. And monuments are erected to him after his death, which of course means that to him *everything* is permitted. No! Such men are not made of flesh and blood, but of bronze! . . .

"The old hag is all rubbish!" he thought heatedly and impetuously. "The old woman is most probably a mistake. She doesn't matter! The old woman is only an illness – I was in a great hurry to step over – I didn't kill a human being – I killed a principle! (291)

When Sonia, who loves Raskolnikov, realizes that he is a murderer, and wonders how he could have done such a thing, he replies that he wanted to become a Napoleon; that's why he killed the old woman. One day, he explains, he asked himself what Napoleon would have done in his place if, instead of a Toulon or Egypt, he had to start his career by murdering some ridiculous old woman – to obtain money for his career, of course. Raskolnikov assures Sonia that he had spent a long time worrying over that question, but eventually concluded that if Napoleon had had no alternative, he would have strangled the old woman without the slightest hesitation. So Raskolnikov hesitated no longer and, following the example of his hero, he murdered her. When he tells Sonia that he has now told her the truth, she cries out, "But what kind of truth is that? Oh, dear God!" But he proceeds to expound his philosophical position, that only he who is firm and strong in mind and spirit can become the master of the people. He who dares much is *right* – that's how the people look at it. He who dismisses with contempt what men regard as sacred, becomes their law-giver; and he who dares more than anyone is more right than anyone. So it has been until now, and so it always shall be. Power is given only to those bold enough to stoop and take it. There is only one thing that matters here – one must have the courage to *dare*. That was his only motive, he tells Sonia, in committing the murder.

Raskolnikov, then, is a man who, as his mother remarked in a letter to him, has succumbed to the modern spirit of godlessness. Under the influence of certain Russian thinkers who themselves had become intoxicated with the proto-Nietzschean ideas of Max Stirner, Raskolnikov developed a theory according to which mankind is divided into masses and higher specimens of men. The latter, owing to their exalted position, are not only above the existing laws and conventions, but they themselves prescribe new laws for the rest of humanity. Impressed with Napoleon and, more generally, with the fact that the strong men of history had paid no attention to individual cases of evil, but stepped over them without giving them a thought, he managed to convince himself that he too was a man of genius. And, though he had some doubts, he continued to persuade himself that he has committed no crime. Unrepentant to the very end, he cannot understand why his action and the rationale behind it strikes everyone around him as so hideous. He acknowledges that he has broken the letter of the law; but he attempts to justify himself by arguing that if he is guilty of any offense, so are those men who, like Napoleon, have caused the death of untold numbers of innocent people. The difference, he tells himself, is that whereas the strong men of history were successful and therefore *right*, he had failed. It is that alone which he considered to be his crime – that he had failed, and therefore had no right to permit himself such a step.

In powerfully dramatic terms Dostoevsky thus called attention to the dangerous moral vacuum created by the doctrine that God is dead – a danger to which Nietzsche gave no consideration.

THE BROTHERS KARAMAZOV (1880)

With this magnificent work of literature, Dostoevsky achieves his greatest triumph not only as a creative artist, but also as a profound and fearless thinker. In what is surely one of the greatest novels the world has ever seen, Dostoevsky integrates idea and character to confront once again the nihilistic ideas of his time. And it is a sure sign of Dostoevsky's greatness as a writer that he makes the case against what he himself stands for much stronger than the case for his own convictions. Father Zossima's pious platitudes are never as convincing as Ivan's "blasphemous" utterances. Ivan, the intellectual protagonist of skepticism and scientific materialism, questions the existence of God and personal immortality. He argues, moreover, that if mankind's belief in immortality were destroyed, not only love but every positive quality on which the continuation of human life depends would dry up at once. There would be nothing immoral then; everything would be permitted. Without the belief in the immortality of the soul, there would be no fear of God, and, hence, no virtue. The source of Ivan's skepticism is his inability to believe that if God exists he would tolerate the world as we know it.

Ivan describes in detail the extreme cruelty of which human beings are capable – so extreme he can't help thinking that if the devil doesn't exist and therefore man has created him, Man has created him in his own image and likeness. And if Ivan cannot accept the world as it is, neither can he accept the Christian doctrine of forgiveness. He has learned the eschatological creed according to which everything in heaven and on earth will one day blend in a hymn of praise and cry aloud: "Thou art just O Lord, for thy ways are revealed!" He cannot accept, however, that on that day the mother will embrace the torturer and murderer of her children. She has no right to forgive him and, besides, too high a price in human suffering and torment has been placed on the ticket for admission to the so-called Kingdom of Heaven. When his devout younger brother responds to Ivan's peroration with "this is rebellion," Ivan continues:

Tell me frankly, I appeal to you – answer me: imagine that it is you yourself who are erecting the edifice of human destiny with the aim of making men happy in the end, of giving them peace and contentment at last, but to do that it is absolutely necessary, and indeed quite inevitable, to torture to death only one tiny creature, the little girl who beat her breast with her little fist, and to found the edifice on her unavenged tears – would you consent to be the architect on those conditions?

And Alyosha replies softly, "No, I wouldn't."[2] Ivan then asks Alyosha whether he is willing to listen to a poetic legend he has written, called "The Grand Inquisitor."

The whole of this poetic legend is the soliloquy of a man who is unable to side either with Christ or with his opposite. Perhaps this man was Dostoevsky himself who, finding no satisfactory answer in Ivan's negations, creates Father Zossima. It is an attempt to answer Ivan's nihilism by means of a Christian synthesis as a guide to life – not in the name of any official Christianity, but rather in the name of Jesus as Dostoevsky understood him. However one may view that synthesis, the fact remains that Dostoevsky delivered a most powerful challenge to the proto-Nietzscheans of his time. Killing God, Dostoevsky reminds his readers, may give rise to the formula that "everything is permitted," a formula which Ivan promised Alyosha he would never repudiate. In Ivan's nightmare, late in the novel, he converses with the devil, and the outlook that emerges could have been written by the author of *Zarathustra*:

> "All that must be destroyed is the idea of God in mankind . . . Once humanity to a man renounces God (and I believe that period, analogous with the geological periods, will come to pass) the whole of the old outlook on life will collapse by itself without cannibalism and, above all, the old morality, too, and a new era will dawn. Men will unite to get everything life can give, but only for joy and happiness in this world alone. Man will be exalted with a spirit of divine, titanic pride, and the man-god will make his appearance. Extending his conquest over nature infinitely every hour by his will and science, man will every hour by that very fact feel so lofty a joy that it will make up for all his old hopes of the joys of heaven. Everyone will know that he is mortal, that there is no resurrection, and he will accept death serenely and proudly like a god. His pride will make him realize that it's no use protesting that life lasts only for a fleeting moment, and he will love his brother without expecting any reward. Love will satisfy only a moment of life, but the very consciousness of its momentary nature will intensify its fire to the same extent as it is now dissipated in the hopes of eternal life beyond the grave . . .
>
> The question now is . . . whether such a period will ever come. If it comes, everything is resolved and mankind will attain its goal. But as, in view of man's inveterate stupidity, it may not be attained even for a thousand years, everyone who is already aware of the truth has a right to carry on as he pleases in accordance with the new principles. In that sense "everything is permitted" to him. What's more, even if that period never comes to pass, and since there is neither God nor immortality anyway, the new man has a right to become a man-god, though he may be the only one in the whole world, and having attained that new rank, he may light-heartedly jump over every barrier of the old moral code of the former man-slave, if he deems it necessary. There is no law for God! Where God stands, there is his place! Where I stand, there will at once be the first place – "everything is permitted" and that's all there is to it!" (763–4)

THE POSSESSED (1871)

On April 23rd 1849, Dostoevsky was arrested as a member of the re-volutionary Petrashevsky Circle consisting of young men interested primarily in the socialist theories of Fourier. Together with twenty other condemned members of the circle he was conveyed to a square in which there stood a scaffold. Minutes before he was to be executed he and his comrades were unbound and informed that the Czar had granted them their lives. The young "rebels" were then sent to Siberia in chains.

However brief Dostoevsky's revolutionary activities may have been, they had a profound impact on his later views. The majority of Dostoevsky's associates in the circle were well-meaning and harmless individuals, carried away by the liberal ideas of 1848 and the utopian French socialism of that era. Such circles of liberal-minded "theorists" and talkers became quite fashionable in Russia after the Napoleonic campaigns. The most notable among them were the Decembrists, whose leaders were executed after the rising of December 1825 – a watershed event in the history of modern Russia. For it was after that event that the political leadership gradually began to pass from the educated nobility to a new and more or less classless body, the intelligentsia, whose formation began in the thirties of the nineteenth century. The intelligentsia was a blend of liberal members of the gentry on the one hand, and of educated commoners on the other. Intoxicated with the latest Western ideas, these Russian intellectuals were determined to rid their country of autocracy and serfdom, although for years they had to content themselves with mere theorizing.

This was the general period in which the proto-Nietzschean ideas of Max Stirner were seized upon in Russia with the greatest enthusiasm. There they formed an important element of the egoist-nihilist-anarchist complex of doctrines. Stirner's influence is best documented in connection with the literary critic, V. G. Belinsky, who was widely regarded as the father of the Russian intelligentsia. Late in life, as Belinsky began to lose interest in literary debates, he turned his attention to contemporary discussions of social and political issues. It was then that he read Stirner, whose *The Ego and His Own* had created a sensation in intellectual circles. Stirner's thesis fascinated Belinsky, and brought about a basic change in the moral credo he had held throughout his life. The new credo became egoism, of course, but an egoism which Belinsky attempted to adapt to Russian political conditions.[3]

It was Belinsky to whom Dostoevsky was largely indebted for his brilliant literary debut. So Dostoevsky was naturally drawn to the famous literary critic during the success of his first novel, *Poor Folk*. But when Belinsky called Dostoevsky's second effort "pathological rubbish," he greatly resented not only the critic's scathing opinion but the critic himself, owing to other personal misunderstandings. Soon Dostoevsky's resentment was directed against everything Belinsky stood for. Nevertheless, Dostoevsky remained for a while under the sway of the radical ideas of the time, despite the fact that those ideas were at odds with the religious atmosphere of his early upbringing. The conflict between his inherited religious propensities and the Enlightenment ideas came to a head during and after his Siberian experiences. His outlook now underwent a total change; and once the change had taken place in him, the Belinsky type of intellectual – a devotee of scientific materialism and rationalism – became his *bête noire*. In his later years, whenever Dostoevsky wrote with bitterness about the Russian radicals, the shadow of Belinsky was in the background.

After the 1861 emancipation of the serfs, hopes were raised among the young liberals that Russia would move in a democratic direction; but reaction soon set in again. The awakened energies turned either to open opposition or to underground activities where they assumed an ominous and destructive character. The extreme left wing of the intelligentsia was particularly resentful. These "nihilists," as they were called, combined philosophical materialism with all sorts of socialist schemes and utopias. Their chief aim was to destroy the old régime in order to clear the ground for something better to come. It was not only their destructive fury, but also their schemes for "something better" that Dostoevsky satirized in his embittered novel about those young people who were "possessed by demons."

In *The Possessed* militant atheism is often represented as the reverse of intense religious faith. It leads to diabolical deceit, crime, and degradation. The real "hero" of the novel is not Peter Verkhovensky, that wicked fool of the underground movement, an unscrupulous apostle of destruction, but the handsome, strong, elegant and mysterious Stavrogin. It is Stavrogin who in his "confessions" relates how he arrived at certain ideas:

> It was then, while sipping my tea and chatting with them about something or other, that, for the first time in my life, I formulated to myself in so many words the idea that I neither know nor feel what evil is. It wasn't

simply that I had lost the feeling of good and evil, but that I felt there was
no such thing as good and evil (I liked that); that it was all a convention;
that I could be free of all convention . . .[4]

About a decade, then, before Nietzsche published the earliest formulations of his mature philosophy, Dostoevsky had placed
proto-Nietzschean ideas in the mouths of his characters, thus reflecting
the fact that such ideas were in the air.

Verkhovensky idolizes Stavrogin, the former Guards officer, and
hopes to turn him into the dark prince of a revolutionary rebellion.
Stavrogin definitely is a "strong type," but his powers lie only in negation. While it is true that Kiriloff, Shatov and others got their nihilistic
ideas from Stavrogin, he is committed to none of them; for him theories
are only a matter of intellectual exercise. This wealthy landowner,
married to a crippled and deranged beggar who sadistically pushes to
suicide the girl-child he had raped, this aristocrat who toys with revolutionary intrigues, is the embodiment of strength without direction.
Stavrogin is equally capable of noble action and of extreme cruelty; he is
attracted by both vice and beauty, degradation and sublimation. For him
everything is permissible as he recognizes no moral code. He belongs,
therefore, in the company of Dostoevsky's other nihilists, those who
challenge not only God, but society and their own consciences as well,
by wilful actions "beyond good and evil." Kiriloff, one of Stavrogin's
"disciples," also proclaims the death of God, and he is possessed by the
idea of a man-god who will transform the world in his own image.
Raskolnikov, Ivan Karamazov, Stavrogin, Kiriloff, and Peter Verkhovensky are all dramatic examples, for Dostoevsky, of the types of
in-dividuals that can emerge as a consequence of philosophical doctrines
which, like Stirner's and Nietzsche's, create a moral vacuum.

Nietzsche is, above all, the philosopher of power who had aimed to
precipitate a crisis by means of his often-repeated assertion that the
world is meaningless and all moral claims are groundless. In that light,
though Nietzsche may have intended his work as a first step in the
process of overcoming nihilism, the impression is unavoidable that his
influence has been such as to deepen the crisis he prophesied. If, as he
believed, "the total nature of the world is . . . to all eternity chaos" (*The
Gay Science*, 109), and if, in addition, the prevailing morality is to be

ignored or repudiated, then why shouldn't individuals come to believe that "everything is permitted," and conduct themselves accordingly? As for Nietzsche's so-called affirmations, they only magnify the moral vacuum, since his "will to power," far from constituting an ordering principle, might rather be regarded – *à la* Hobbes – as a source of continuing disorder.

NOTES

1   *Crime and Punishment*, tr. David Magarshack (Penguin Books, London, 1951), p. 208. All page references to this book are hereafter cited in parentheses immediately following the quoted passage.
2   *The Brothers Karamazov*, tr. David Magarshack (Penguin Books, London, 1982), pp. 287–8. All page references to this volume are hereafter cited in parentheses immediately following the quoted passage.
3   See John Carroll's Introduction to Max Stirner, *The Ego and His Own*, tr. Steven T. Byington (Jonathan Cape, London, 1971), p. 28.
4   *The Possessed*, tr. Andrew R. MacAndrew (Signet Classics, New York, 1980), p. 426. See Marc Slonim's Afterword for an illuminating commentary on this novel.

EPILOGUE

Thus Spoke the Prophets of Social Justice

Nietzsche's admiration for the "master morality" carries with it a profound contempt for the "rabble" and for democracy. His attitude in this regard was a matter of taste, not reasoned argument. As we noted earlier, however, he also greatly admired the Old Testament, where "there are individuals, things, and speeches in so grand a style that Greek and Indian literature have nothing to compare with it (*Beyond Good and Evil*, section 52). Moreover, he recognized the "this-worldly" character of biblical Judaism. That being the case, it should be interesting to listen to the classical spokesmen for the "slave morality," the prophets of social justice, and to counterpose their message to that of Zarathustra and his creator.

The inversion of values in ancient Israel, as we have seen, began with the slave experience of a mixed multitude of Semitic nobodies who rebelled against the most powerful "man-god" of the time, the Pharaoh of Egypt. The Exodus, as a successful liberation from servitude, became thereafter the archetypal experience by which human relations would be evaluated where freedom and justice were concerned. It was Moses, most likely, who first intuited an almighty God who had heard the cry of an oppressed people and who led them out of bondage, thus demonstrating his power and justice. That this formless, invisible God was greater than any and all of the worldly powers became apparent to the people in his ability to make all forces, political or natural, serve him. From the time of that original experience, the Israelites relied for their knowledge of

God's will on what actually happened in history. The Israelites had thus trained their eye from earliest times to take human events seriously, because in them one learned more clearly than anywhere else what God willed and what he was about.

The genuinely new element in classical prophecy was the idea that Israel, and indeed all humanity, will be judged by its social conduct. At about the time that Hesiod, in ancient Greece, spoke out against the "crooked" magistrates, Amos denounced the extreme oppression and exploitation of the poor by the privileged classes. Like the other prophets of the eighth century BC, Amos directs his anger against the wealthy and powerful because it is they who "turn justice to wormwood, and cast righteousness to the ground" (Amos 5:7). They trample upon the poor and turn aside the needy at the gate. Those who lie on beds of ivory are accused of "swallowing the needy and destroying the poor of the land," of "falsifying the balances," of buying the "poor for silver, and the needy for a pair of shoes" (8:4–7). Those are some of the injustices that Amos condemns in the strongest possible terms. Also new with Amos is the corollary that the cult, with its sacrificial ceremonies, has no religious validity in and of itself. He repudiates everything associated with the cult if it is unaccompanied by fairness, honesty and decency in human relations (5:21–25). Twice Amos harks back to the fact that the Lord brought this people out of the land of Egypt (2:10, 9:7), implying thereby that this formerly servile people, above all, ought to know how to behave towards their fellow human beings.

Hosea also reminds his listeners that they are the people who were brought up from the land of Egypt (Hosea 12:10); and like Amos, he denies any independent validity to the cult: "For I desire mercy, and not sacrifice, and the knowledge of God rather than burnt offerings" (6:6). Equally important for our purposes, Hosea denounces the princes and noble lords for putting their trust in might rather than right:

> Ye have plowed wickedness, ye have reaped iniquity, ye have eaten the fruit of lies; for thou didst trust in thy way, in the multitude of thy mighty men. Therefore shall a tumult arise among thy hosts, and all thy fortresses shall be spoiled. (10:13–14)

> Because of the wickedness of their doings I will drive them out of My house; I will love them no more, all their princes are rebellious. (9:15)

And I shall break the bow and the sword and the battle out of the land. (2:20)

For Israel hath forgotten his Maker, and builded palaces, and Judah hath multiplied fortified cities. (8:14)

Isaiah, the most universalist of the prophets, also demands justice: "relieve the oppressed, judge the fatherless, plead for the widow" (Isaiah 1:17). Like Amos and Hosea, he indicts the privileged elders and princes:

It is ye that have eaten up the vineyard; the spoil of the poor is in your houses; what mean ye that ye crush My people, and grind the face of the poor? (3:14)

Evidently there was a kind of "Enclosure" movement going on at the time, in which the princes and the large landowners were annexing the land of the poorer peasants:

Woe unto them that join house to house, that lay field to field, till there be no room, and ye be made to dwell alone in the midst of the land. (5:18)

Already then there were haughty princes who conducted themselves as though they were "beyond good and evil." Isaiah therefore castigates the princes for their *hubris* – for fancying themselves to be high and mighty, for building fortified walls with lofty towers and putting their trust in their own might (2:16–17). "Woe unto them," he declares,

that call evil good and good evil . . . Woe unto them that are wise in their own eyes . . . Woe unto them that are mighty to drink wine, and men of strength to mingle strong drink; that justify the wicked for a reward, and take away the righteousness of the righteous from him! (5:20–5:23)

And, of course, Isaiah's most powerful repudiation of the princes and their "master morality" is found in the prophet's best known and universalist vision:

And it shall come to pass in the end of days, that the mountain of the Lord's house shall be established as the top of the mountains, And shall be exalted above the hills; And all nations shall flow unto it. . . . And He shall judge between the nations, and shall decide for many peoples; And they shall beat their swords into plowshares, and their spears into pruning hooks; nation shall not lift up sword against nation, neither shall they learn war any more. (2:2–4)

The prophet Micah likewise rebuked those who behaved as if nothing is forbidden them:

Woe to them that devise iniquity . . . because it is in the power of their hand. And they covet fields and seize them; And houses, and take them away; Thus they oppress a man and his house. (Micah 2:1–2)

It was the masters and rulers who committed the worst offenses:

Hear, I pray you, ye heads of Jacob, And rulers of the house of Israel: Is it not for you to know justice? who hate the good and love the evil; who rob their skin from off them, and their flesh from off their bones; . . . Then shall they [the rulers] cry unto the Lord, but he will not answer them; Yea, He will hide His face from them at that time, according as they have wrought evil in their doings. (3:1–4)

In a passage that will remind us of Hesiod's denunciation of the crooked magistrates, Micah declares,

The prince asketh, and the judge is ready for a reward. (7:3)

We have limited ourselves to the prophets of the eighth century BC, but perhaps that is enough to make the point we are after: they are the great originators and spokesmen for what Nietzsche called the "slave morality," the morality demanding that the weak be protected from the strong – the morality Nietzsche has sought to discredit. In his rejection of the ethical principles implicit in this "slave morality" he tends to commit the error called, in philosophy, the "genetic fallacy." This refers to the erroneous assumption that the social origin of a value or idea has necessary implications for its validity. However, the fact that the ethical principles enunciated by the prophets had emerged out of the slave experience of a people has no necessary implications for the validity of those principles. Nietzsche therefore fails to meet the exponents of these principles on their own ground. He never grapples with these principles as such, by comparing them objectively with "master morality" tenets and asking which are the sounder guidelines for social life. He merely prefers and endorses the "master morality" as a matter of taste, as we have seen. If, however, the quality of life in the human condition is at issue, all one needs to do is to compare Zarathustra's teachings with those of Amos, Hosea, Isaiah, and Micah to see that it is no contest.

We know what Nietzsche was against; but what did he affirm? If there are any real affirmations in Nietzsche's writings, they may be summed up in his three key ideas: the "will to power," the yearning for the "superman," and the "eternal recurrence." But these ideas are beset with problems, defects, and ambiguities. Take his notion of the "will to power." The intended meaning of this phrase is not entirely clear. There are aphorisms in which the phrase seems to refer, simply, to the will to life – to the affirmation of the natural and instinctual side of the human being. However, Nietzsche sometimes speaks as if it is the self-mastery and sublimation of the instincts that is to be admired, while at other times he speaks as if it is the total freedom and spontaneity of the instincts that he advocates – an impossible condition in any case, since it implies, in Freudian terms, that the "id" be totally unharnessed. The phrase "will to power" often also has the same meaning as it has for Hobbes; and yet, as we have seen, Nietzsche fails in such instances to take into account the Hobbesian proposition that conflict is bound to ensue when two or more individuals will or desire the same future apparent good, which nevertheless they cannot both have.

As for the yearning for the advent of the "superman" and the notion of the "eternal recurrence," both of these ideas seem to stand in a contradictory relationship to the basic assumption of his philosophy, that the universe is in a state of chaos. For if the universe is chaotic, what possible effect can proclamations of the superman have? And if the universe is lacking in any principle of order, what ground does Nietzsche have for his belief in the "eternal recurrence," which implies an order of a cyclical kind? In any event, it is certain that these "affirmations" could never fill the void Nietzsche helped create with his negations. Nor, certainly, did he expect that they would, since he specialized in philosophizing with a hammer.

For Nietzsche, as we have seen, there are no first principles, no *archai*. It is an illusion originally fostered by Socrates that through reason and dialogue we can not only know reality but correct it in accordance with fundamental truths. Thought, for Nietzsche, can no longer pretend to operate in the realm of truth and validity claims. Nietzsche thus leaves us with his own peculiar, beyond-good-and-evil aestheticism which enthrones *taste* as the sole means of arbitrating between values.

Now it is probably true that many of us come away from Socrates' encounter with the proto-Nietzscheans not wholly convinced that he has decisively defeated them, just as we may have nagging doubts about the claims of contemporary guardians of reason who propose that the demand for freedom and the practice of dialogical reasonableness possess a transcending power. The experience of the twentieth century lends but little support for such claims. Even among the professional philosophers themselves there is no agreement on the nature of the arguments advanced in any of the "canonical" texts; nor, certainly, is there a consensus on who has advanced the best argument. Indeed, there is an extreme view in contemporary philosophy that argument is a ridiculous philosophical obsession. Argument, it is said, is nothing more than rhetoric, nothing more than attractive vocabularies designed to persuade. For proponents of this view there is no irrefutable ground for the belief that, say, cruelty or genocide are horrible and wrong. The "universal truths" by which such conduct has traditionally been condemned are regarded as mere platitudes of an entrenched vocabulary. Moreover, there are no neutral criteria by which to evaluate rival vocabularies.

Critics of this extreme view concede that definitive foundational justifications cannot be given in support of one's beliefs; only historically-contingent, fallible reasons can be offered. From Nietzsche's aesthetic standpoint, however, even historically-contingent, fallible reasons are not to be admitted. His "bad" aestheticism prevents him from considering the likelihood that his yearning for the "higher specimens" and his contempt for the multitude are simply manifestations of his snobbish, anti-human prejudices. And insofar as he draws from nature the inference that the higher or stronger human types must segregate themselves from the "herd," he flies in the face of the evidence. For in the animal world, as we have seen, the stronger types provide a leadership which enhances the adaptability of a species and redounds to the benefit of all. In their *intra*-species behavior, Darwin assures us, animals follow the Golden Rule. Animals have sufficient cognitive capacity to recognize the mutual benefit to be gained from following that Rule. The human species as a whole, in contrast, has yet to emulate the animals in that respect, the major reason being that humanity, unlike other species, is divided against itself.

We need, therefore, to refer once again to the ninety-sixth aphorism in *The Dawn*, where Nietzsche, "in accordance with the commandments of reason," had called for a power that would mediate between the nations, between the classes, between rich and poor, between rulers and subjects, between warlike and peaceful peoples. It is to be regretted that he applied his powerful intellect not to this project but to the one with which we are familiar, and which contributed nothing to the realization of the ideal enunciated in that aphorism. In a word, he should have remembered that it was not only Socrates but also Isaiah who, centuries earlier, had said: "Come now, and let us reason together" (Isaiah 1:18). Nietzsche should have known that the Dionysian passion of his protagonist, Zarathustra, could only be blind and destructive without the guidance of the exemplary reasoning of Isaiah and the other prophets of social justice.

Index

Index